海派时尚

上海设计之都公共服务平台 认定专业平台

海派时尚 STYLE SHANGHAI

STYLE SHANGHAI

2015春夏海派时尚流行趋势
2015 SPRING/SUMMER STYLE SHANGHAI FASHION TREND

配饰篇
ACCESSORY

海派时尚流行趋势研究中心 著

东华大学 出版社

[配饰篇]

编辑委员会
名誉主任：厉无畏
顾问：徐明稚、Christopher Breward、席时平
主任：刘春红
执行主任：刘晓刚
副主任：刘健、贺寿昌、杲云、朱勇
委员：林艺、庄培、吕苏宁、邵峰、蒋智威、徐晶
卞向阳、吴翔、彭波、汪芳、钟宏
统筹：李峻、曹霄洁、吴亮、傅白璐
编排：刘清芸、胡弘毅、周吉、范乃文

主创人员：

俞英、田玉晶、赵舟旭、王瑜婷、吴凯立、汝海洋、刘文彬、周姝君、戴伟豪、汪子昀、严克、龙城、郑迪斐、
韩宇静、王依宣、周璇、吴琼、杨景裕、覃羽、李梦晓、易丽思、刘蔚琪、李敏航、贺聪

鸣谢：

上海雅氏鞋业有限公司
红谷集团
帽仕汇
浙江美联工贸有限公司

支持单位：

上海市经济和信息化委员会
上海市文化创意产业推进领导小组办公室
上海市长宁区人民政府
东华大学
上海智富企业发展（集团）有限公司
司色艾印染科技（上海）有限公司
上海国际服装服饰中心

翻译：

Translation Graciously Provided by:The Odyssey Network
Steve Zades （Donghua University Consultant Professor for Postgraduates,USA）
Zheng Yuan Mao

2015春夏海派时尚流行趋势主题
2015 SPRING/SUMMER STYLE SHANGHAI FASHION TREND THEMES

梦 DREAMS

电影是造梦的艺术,在国产电影走向世界的进程中,梦想的轮廓也愈加圆满。作为海派时尚独特表达的海派电影,是海派时尚设计的重要资源。在海派电影摇曳的光影中,海派时尚的脉络也愈发清晰。

Film is the art of dreaming, in the process of Chinese films stepping to the world, the outline of dream is becoming even more complete. As a unique fashion expression, Shanghai movie is an important resource for Style Shanghai fashion design. Shanghai movie swaying in light and shadow, the context of Style Shanghai is also increasingly clear.

海上星梦 [海派经典风格] STAR DREAMS OVER THE SEA [SHANGHAI CLASSICAL STYLE]

近代海派电影的繁荣造就了摩登生活方式的汇集,创新设计与怀旧经典相互交融。光影流动的时尚教科书蛊惑着沪上各阶层女性,她们很快成为电影所传诉的风尚追逐者和推波助澜者。

Modern Shanghai Movies created a modern lifestyle collection, innovative design and retro classics mingled. Light flowing fashion textbook charms the Shanghai women, and they soon became the movie fashion chaser.

优活之梦 [海派自然风格] LIVING THE DREAM [SHANGHAI NATURAL STYLE]

弄堂生活是水泥森林中的桃花源。在申城日渐国际化的今天,反映海派市井日常生活的影视剧持续风靡,也反映了当代人对慢活、乐活、优活理念的秉持。来自家庭手工式、生态环保的弄堂作品是现代都市人心中最温暖的城市记忆。

Alley life is paradise in concrete jungle. While Shanghai is becoming increasingly international, the movies of Shanghai daily life continues to be popular, reflecting contemporary people's persistence for slow living, lohas, and quality of life. Alley inspired handmade style from family, eco-friendly design work is the warmest memories of the modern city.

时空筑梦 [海派未来风格] REALISING DREAMS THROUGH SPACE [SHANGHAI FUTURISTIC STYLE]

高科技从思维、行为到审美的各个方面对人们的衣食住行产生影响。炫目的科幻大片以强烈的视觉冲击力牢牢吸引着城市年轻群体的目光。强规则几何图形、固化的符号、机械生物等是人们对未来生活的深思与肆意畅想。

High-tech is having impact on people's daily life, from thinking, behavior to the aesthetic aspects. Stunning sci-fi blockbuster with strong visual impact firmly attracted the attention of young urbanites. Strong geometry, curing symbols, mechanical and biological life are composing thinkings about the future.

多彩享梦 [海派都市风格] ENJOY COLORFUL DREAMS [SHANGHAI URBAN STYLE]

坚持个人理想、追求幸福与欢乐的个人价值的都市人,用个性打破思维的界限,在叛逆和玩乐中创意。多彩的感官享受,幽默、趣味和诙谐都让我们的"快乐崇拜"不停歇。

Adhere to personal dream, with pursuit of happiness and joy, individual thinkings to break the boundaries of creativity and play in the rebellion. Colorful sensuality, humor, fun and witty, makes our "Happy worship" never stop.

Preface

While changing with times, fashion trends have always been strongly influenced by social and cultural movements. Since successfully launching the first issue of "Style Shanghai", our team have continued to trace and further study the origins of Shanghai style from different aspects including movies, literature, music, drama and etc.

Film, an important part of fashion culture, is not only what we use to make dream come true in virtual world, but also a popular resource of design inspiration. Shanghai Style movies hold an irreplaceable position in the Chinese film history. Since movie was first publicly introduced to Chinese people in Shanghai Tianhua Tea House in 1897, this industry has developed a lot from unforgettable 1930s movie stars, such as Die Hu and Lingyu Ruan, famous for their incredible beauty and talent to thousands of films of different genres, which have inspired our designers. This book will tell you many design stories behind our work with Shanghai Style movie as a main theme and other 4 subthemes, classical style, natural style, futuristic style and urban style.

For your convenience, we have tailored this book with 3 chapters, Inspiration, Apparel and Accessories. Inspiration includes articles on Shanghai style culture, fabric and graphics; Apparel is divided into men's and women's; Accessories covers footwear, bag, hat, jewelry and etc.

As a project to foster Shanghai culture and innovation industries and a main product of Greater Donghua service platform for innovation industry development, this book's publication has received a lot of support from Shanghai Municipal Commission of Economy and Information, Bureau of Changning District, Donghua University and other sectors of society. Our design team is mainly composed of experts from Donghua University, bringing together artists, designers, entrepreneurs and international friends. With almost 100 contributors' 6-month collaboration and continuous efforts, we represent you this second issue of Style Shanghai Fashion Trend.

Fashion tells about attitude, culture, spirit and way of living. Chinese fashion dreams are about having our own styles, trends, creative designs and fashion identity on world stage. In order to realize those dreams, we will persist in originality, inherit the essence from the tradition, popularize Style Shanghai fashion and build a more beautiful China!

Style Shanghai Fashion Trend Research Center

序

　　流行趋势植根于时代，时尚潮流溯源于文化，自"海派时尚"创刊号成功发布后，创作团队就海派时尚的文化与时代的交汇进行了追根溯源和深入探讨。透过海派电影、海派文学、海派音乐、海派戏剧……发现海派时尚的文化之源。

　　电影，是时尚文化重要的组成部分，是塑造精神之梦的艺术体裁，更是时尚设计师们热衷的灵感源泉。海派电影，在中国电影发展历程中有着不可替代的重要地位。自1897年于上海天华茶园电影首次传入中国，20世纪30年代风华绝伦的电影明星胡蝶、阮玲玉等的出现，到今天海纳百川的电影新纪元，都成为设计师们枝繁叶茂的灵感花园。2015年春夏海派时尚流行趋势以海派电影为主线，分别从海派经典风格、海派自然风格、海派未来风格、海派都市风格等方面进行演绎。

　　为了方便时尚业者按图索骥，分类检索，2015年春夏海派时尚流行趋势分为三大篇章，分别为灵感篇、服装篇和配饰篇。其中，灵感篇包括海派文化、面料、图形章节，提供海派时尚趋势设计的灵感和素材；服装篇包括海派男装和女装章节；配饰篇包括海派鞋履、箱包、帽饰、首饰等章节。

　　作为上海市文化创意产业扶植项目、环东华时尚创意产业服务平台的重要内容，本书的出版得到了上海市经信委、长宁区政府、东华大学各级领导和社会各界的广泛支持。团队以东华大学设计学科专家为主体，汇集了艺术家、设计师、企业家、国际友人，团队成员近百名，共同迸发出创意火花，循序渐进，数易其稿，历时半年完成。

　　时尚，是当代人所崇尚的一种态度、一种文化、一种精神，更是一种生活方式。时尚中国梦，是中国人拥有自己的原创风格、流行趋势、创意设计，在世界时尚界有属于自己的时尚话语权。坚持原创驱动，打造海派时尚，建设美丽中国！

海派时尚流行趋势研究中心

目录 CONTENTS

序 PREFACE	4
2015 春夏海派时尚流行趋势主题：梦 2015 SPRING/SUMMER STYLE SHANGHAI FASHION TREND THEMES: DREAMS	6
鞋履 FOOTWEAR	8
海上星梦 STAR DREAMS OVER THE SEA	10
优活之梦 LIVING THE DREAM	22
时空筑梦 REALISING DREAMS THROUGH SPACE	34
多彩享梦 ENJOY COLORFUL DREAMS	50
箱包 BAG&SUITCASE	68
海上星梦 STAR DREAMS OVER THE SEA	70
优活之梦 LIVING THE DREAM	84
时空筑梦 REALISING DREAMS THROUGH SPACE	98
多彩享梦 ENJOY COLORFUL DREAMS	112
帽子、首饰及其他 HAT, JEWELRY&OTHERS	128
海上星梦 STAR DREAMS OVER THE SEA	130
优活之梦 LIVING THE DREAM	140
时空筑梦 REALISING DREAMS THROUGH SPACE	150
多彩享梦 ENJOY COLORFUL DREAMS	160
2015 春夏海派时尚流行要素汇编·配饰篇 2015 SPRING/SUMMER STYLE SHANGHAI FASHION ELEMENTS COLLECTION·ACCESSORY	170

2015 SPRING/SUMMER STYLE SHANGHAI FASHION TREND

经典
CLASSICAL

西方
WEST

东方
EAST

未来
FUTURE

海上星梦
STAR DREAMS OVER THE SEA
海派经典风格
SHANGHAI CLASSICAL STYLE

优活之梦
LIVING THE DREAM
海派自然风格
SHANGHAI NATURAL STYLE

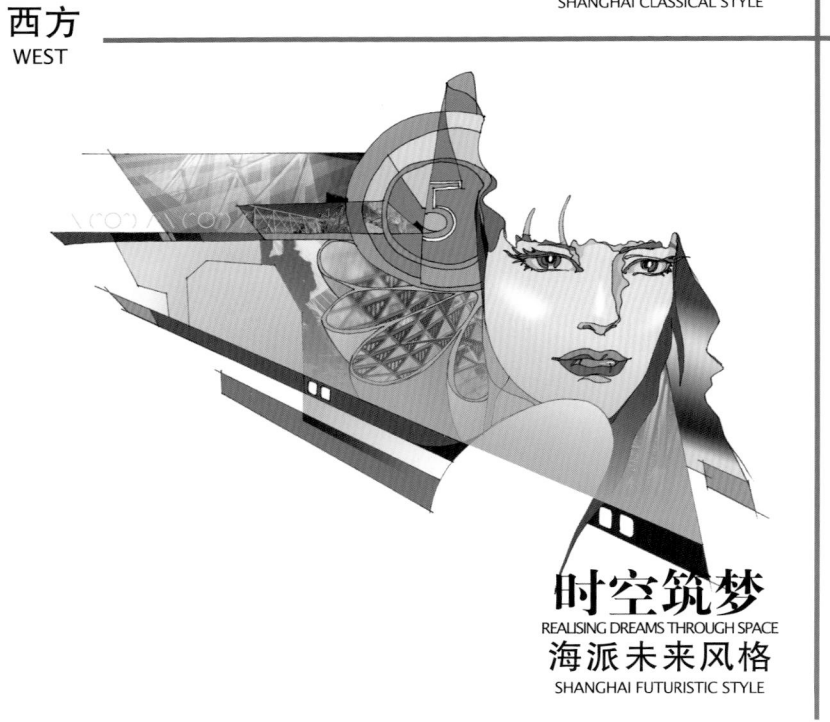

时空筑梦
REALISING DREAMS THROUGH SPACE
海派未来风格
SHANGHAI FUTURISTIC STYLE

多彩享梦
ENJOY COLORFUL DREAMS
海派都市风格
SHANGHAI URBAN STYLE

8   www.style.sh.cn

鞋履

FOOTWEAR

海上星梦
STAR DREAMS OVER THE SEA
优活之梦
LIVING THE DREAM
时空筑梦
REALISING DREAMS THROUGH SPACE
多彩享梦
ENJOY COLORFUL DREAMS

S 鞋履 2015 春夏海派时尚流行趋势

海上星梦
STAR DREAMS OVER THE SEA

夜上海的霓虹色彩散发着棕色的经典气息，细高跟鞋、牛津鞋、乐福鞋等的设计款式，结合各种经典花纹面料和牛皮材质，不停上演着对繁华老上海的热爱和怀旧。

Neon lights in Shanghai night exude classic feelings. Stiletto, Oxford shoes, Loafers and other design styles together with printed fabrics and leather arouse people's interests and nostalgia of old Shanghai.

FOOTWEAR

2015 SPRING/SUMMER STYLE SHANGHAI FASHION TREND

海派鞋履流行趋势

2015 SPRING/SUMMER STYLE SHANGHAI FASHION TREND / FOOTWEAR

关键要素 | KEY POINTS

繁华时代
FLOURISHING ERA

精细品质
EXQUISITE QUALITY

新贵气质
NOBLE TEMPERAMENT

创新传统
INNOVATING TRADITIONS

青釉灰

青砖灰

杏仁黄

皮革黄

琥珀橙

酸枝红

紫檀棕

瓷瓦黑

铁艺灰

青瓷灰

S 鞋履 2015 春夏海派时尚流行趋势

海上星梦
STAR DREAMS OVER THE SEA

花样年华海报

静安寺夜景

墙面装饰

气球等

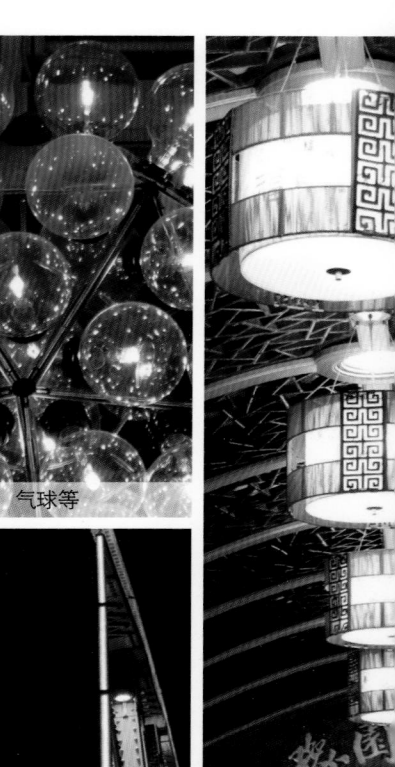

豫园老街的灯

老上海海报

外白渡桥

设计灵感

经典的乐福鞋和牛津鞋，在鞋面上运用了花纹面料，打破了一般传统鞋面沉闷的感觉，颜色上符合春夏比较大胆靓丽的习惯，营造出繁华经典的感觉。

We use patterned fabrics on the surface of these Classic Loafers and Oxford, breaking dullness of traditional ones. We also use bold colors for Spring/Summer collection, creating a bustling but classic feeling.

印花面料

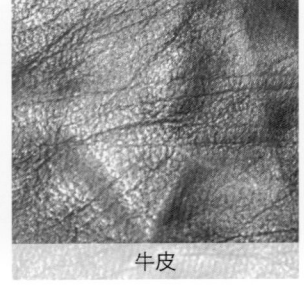

牛皮

编织牛皮

12　www.style.sh.cn

2015 SPRING/SUMMER STYLE SHANGHAI FASHION TREND / FOOTWEAR

关键要素 | 繁华时代
KEY POINTS | FLOURISHING ERA

铁艺灰

雷雨灰

胭脂红

麦糖黄

杏仁黄

细高跟配上雾面金葱或印花鞋面，整体上体现繁华之感。
Flower printed vamp together with a foggy glitter printed stiletto makes this pair of shoes a luxury.

扁型打蜡鞋带

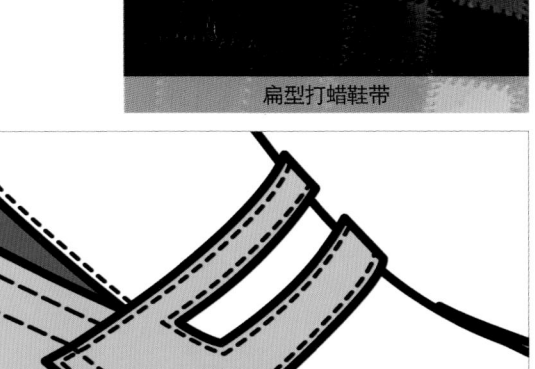
乐福鞋头细节

皮鞋绣花细节

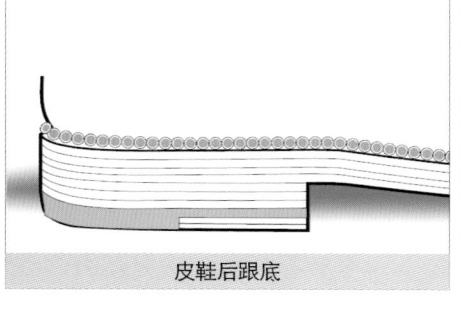
皮鞋后跟底

重复图案

植绒面料

蕾丝花纹

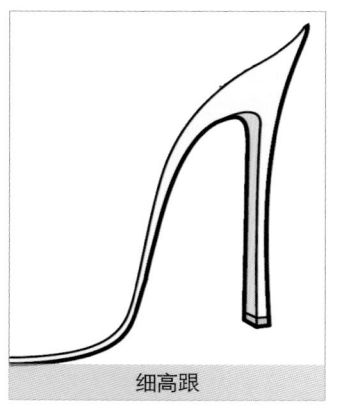
细高跟

在经典鞋型的基础上，咖啡色配合白色底的印花面料。
Sand and white brogue details dance across a classic style in the same hue.

蕾丝面料

印花布

金沙布

雾面金葱

印花面料

13

 鞋履 2015 春夏海派时尚流行趋势

海上星梦
STAR DREAMS OVER THE SEA

露天下午茶

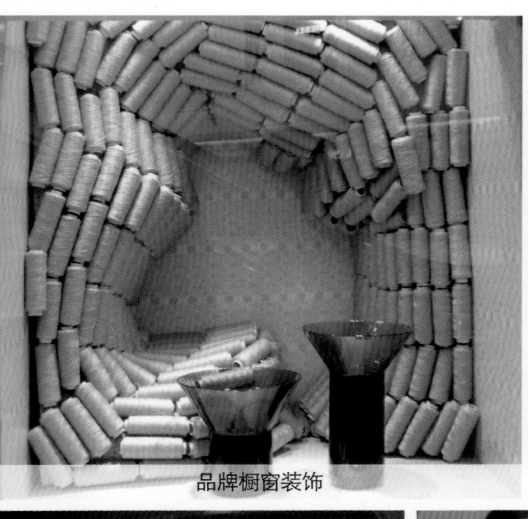

品牌橱窗装饰

橱窗海报

新天地露天酒吧

立体剪纸

上海老怀表

精致灯饰

设计灵感

用各种牛皮皮料结合经典款型打造的精细经典男鞋，以反差色的包边和手缝线的细节为主，突出精细这一特质。

Exquisite classic men's leather shoes with contrasting lining and stitching.

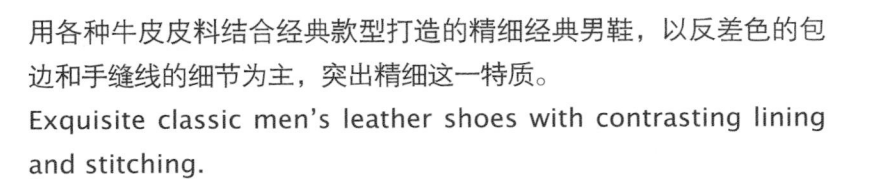

雾面牛皮

细纹理牛皮

咖啡色裂纹牛皮

14　www.style.sh.cn

2015 SPRING/SUMMER STYLE SHANGHAI FASHION TREND / **FOOTWEAR**

关 键 要 素 | 精细品质
KEY POINTS | EXQUISITE QUALITY

淡奶咖

皮革黄

瓦罐棕

豆浆白

曜石黑

粗跟的包牛皮鞋跟，白色牛皮面料上配合金色的包边。
Stacked heel with white bull leather upper and gold lining.

皮鞋鞋带

经典沿条缝线

珍珠

粗手缝包边装饰

手缝线

鞋带头

包边处理

以黑白经典色为主的细高跟女鞋，后跟配上珍珠或细钻，更好的诠释精细这一点。
Stiletto represents exquisiteness. Classic black and white with some decorations of pearl and diamonds better conveys this.

细高跟

皱面牛皮

扁型打蜡鞋带

深色各色皮料

荔枝纹牛皮

双针缝线

15

  鞋履 2015 春夏海派时尚流行趋势

海上星梦
STAR DREAMS OVER THE SEA

蓬蓬裙创意雕塑

酒店大堂门前雕花

欧式经典花纹

手工磨咖啡机

大理石雕花

经典吊灯

放映机

设计说明

千鸟格和蕾丝面料，搭配细高跟和金属中跟，经典款型搭配欧式印花面料，有不一样的感觉，但同样体现新贵感。

Houndstooth-printed lace fabric with stiletto or metallic pump, classic style with European-style fabric, gives us a brand-new upstart sense.

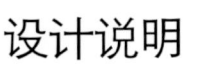

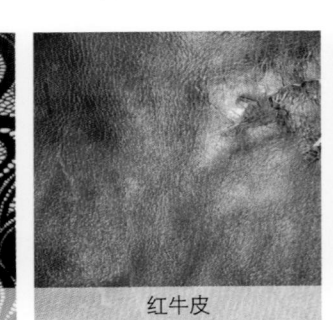

蕾丝面料

红牛皮

千鸟格

16 www.style.sh.cn

2015 SPRING/SUMMER STYLE SHANGHAI FASHION TREND / FOOTWEAR

关 键 要 素 | 新贵气质
KEY POINTS | NOBLE TEMPERAMENT

紫檀棕

铁艺灰

青烟蓝

浪花白

青釉灰

经典男士绑带牛津鞋，配合暗色印花同色系拼接。
Classic lace-up Oxford with dark shade brogues.

皮质流苏

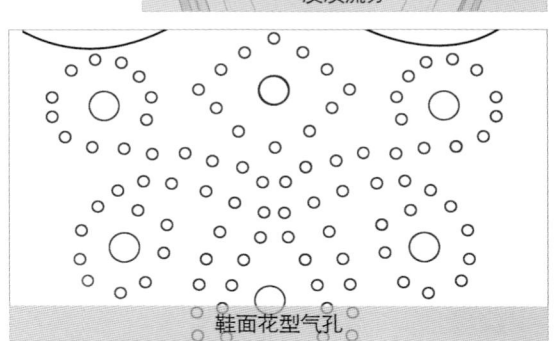
鞋面花型气孔

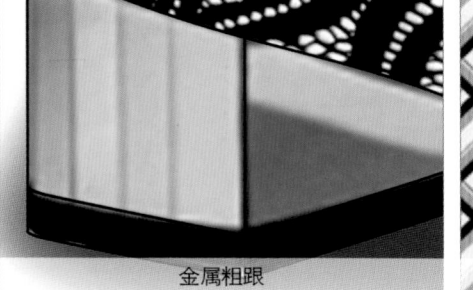
金属粗跟

鞋面印花细节

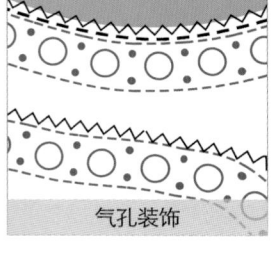
气孔装饰

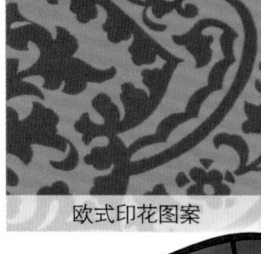
欧式印花图案

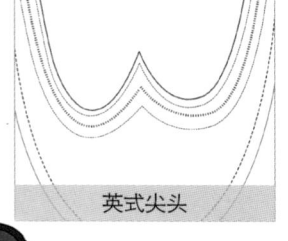
英式尖头

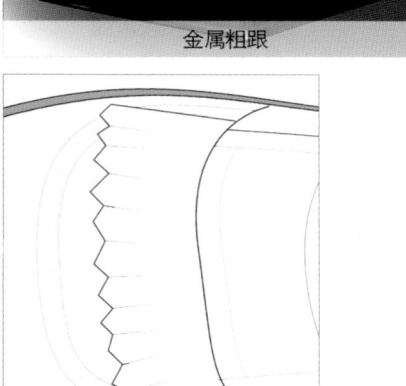
短流苏

新贵男士鞋以细包边和前绑带打结流苏为细节，鞋面为印花面料。
Upstart men's shoes with thin lining and front straps with tassel details, leather uppers and brogues.

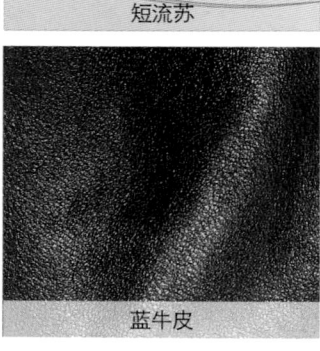
蓝牛皮

皮质鞋带

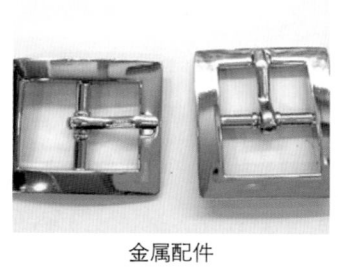
金属配件

印花面料

尼龙鞋带

  鞋履 2015 春夏海派时尚流行趋势

海上星梦
STAR DREAMS OVER THE SEA

外滩夜景

夜间街景

圣诞节外墙装饰

欧式装饰盘

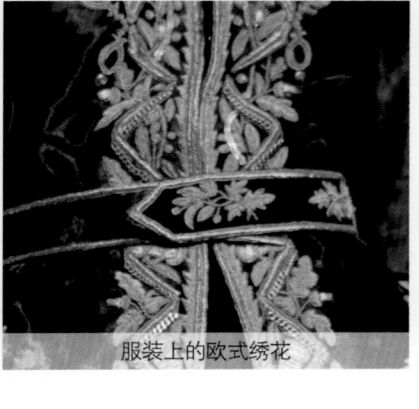

服装上的欧式绣花

餐具瓷器

各色灯具

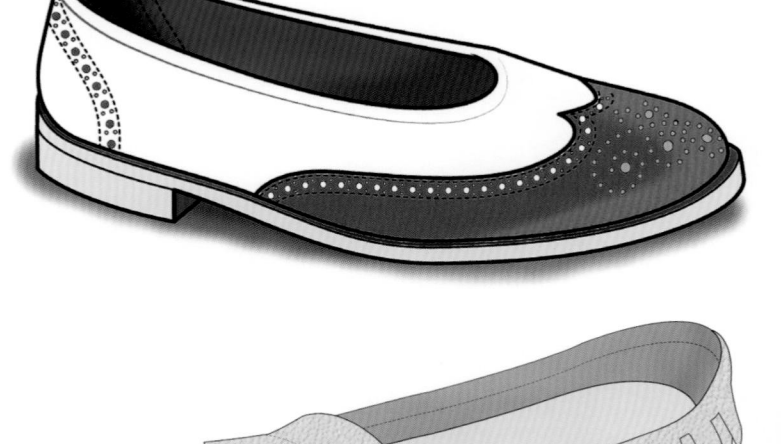

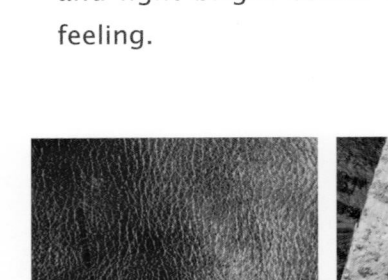

设计灵感

新贵童鞋款式上以牛津鞋为主，尖头短流苏和鞋面气孔为主要细节，在颜色方面以淡色亮色体现春夏感。

Child shoes with Oxford wingtip, short tassel and air holes and light bright colors in consistent with Spring/Summer feeling.

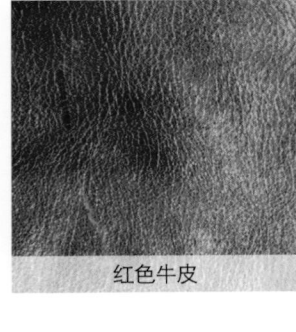

红色牛皮

植绒印花面料

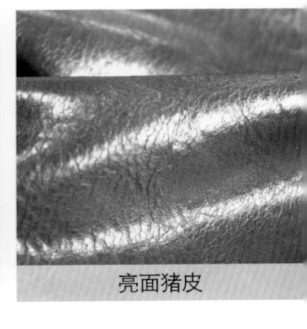

亮面猪皮

18　www.style.sh.cn

2015 SPRING/SUMMER STYLE SHANGHAI FASHION TREND / FOOTWEAR

关键要素 | 新贵气质
KEY POINTS | NOBLE TEMPERAMENT

枫叶红

酸枝红

夜空蓝

乌梅黑

烟云灰

厚底休闲运动鞋，颜色反差大。
Athletic shoes with rubber sole and contrasting colors.

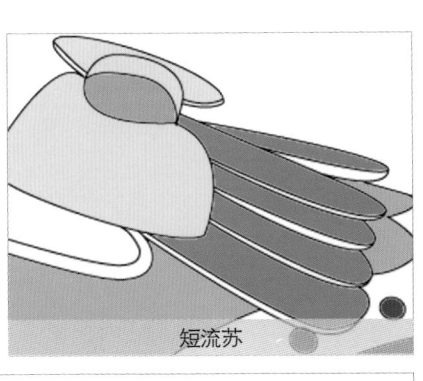

短流苏

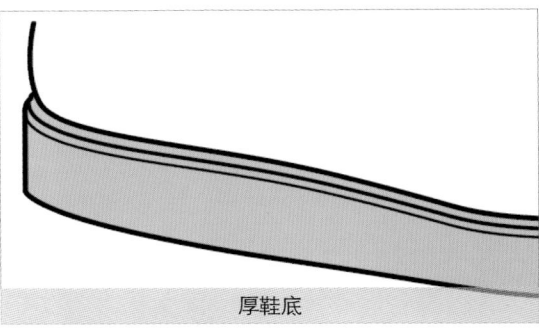

厚鞋底

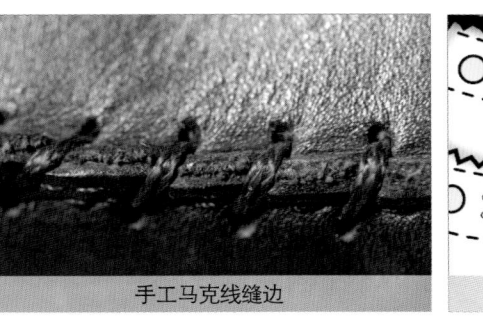

手工马克线缝边

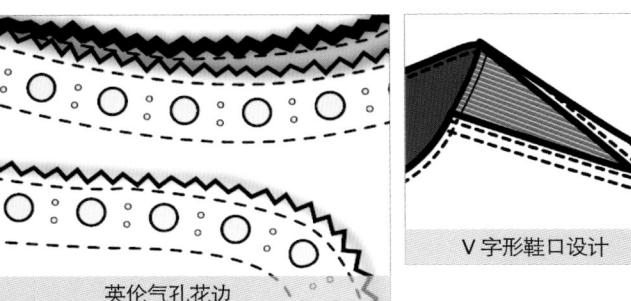

英伦气孔花边

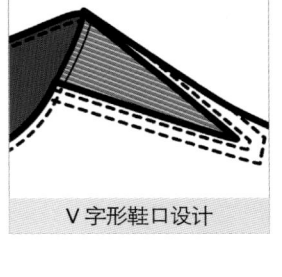

V字形鞋口设计

千鸟格细节

花形气孔

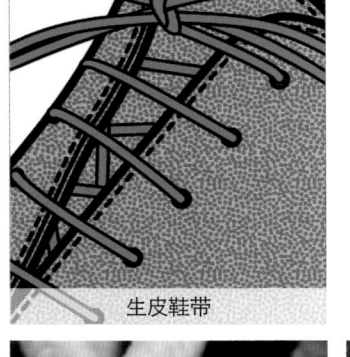

生皮鞋带

同系列的款型，搭配不同颜色的拼色，有不同的感觉。
Shoes from same collection. Different color matches to choose from giving you different feelings.

运动鞋带

拼接面料

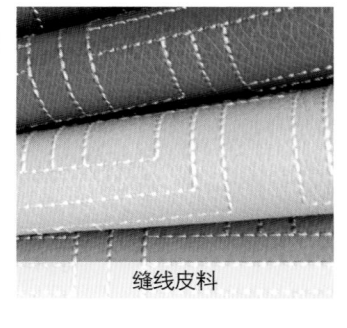

缝线皮料

白色皮料

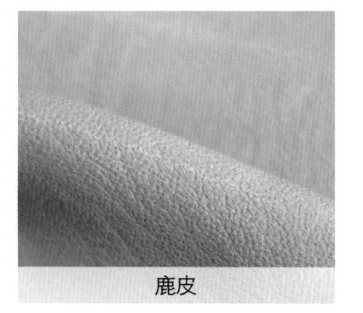

鹿皮

鞋履

2015 春夏海派时尚流行趋势

海上星梦
STAR DREAMS OVER THE SEA

创新报纸 "服装"

创新展示

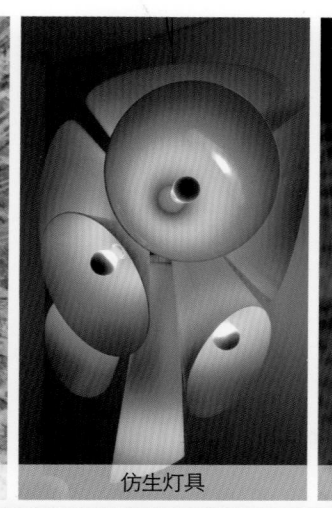

仿生灯具

新型灯具

创意面料展

新材料展

视觉展示面料

设计灵感

不同材料的拼接，透明 PVC 和运动绑带加上印花鞋头和草编与皮质加绣花面料的相结合，运用的是经典鞋款，营造经典创新之感。

Different material matches such as transparent PVC and athletic lace-up with printed wingtip, or braids and leather with embroidery on classic shoes is an example of "Classic plus Innovation".

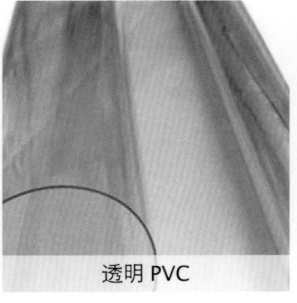

透明 PVC

印花面料

草编材料

2015 SPRING/SUMMER STYLE SHANGHAI FASHION TREND / **FOOTWEAR**

关键要素 | 创新传统
KEY POINTS | INNOVATED TRADITIONS

豆沙红

雷雨灰

胭脂红

浪花白

瓦罐棕

软木底搭配经典牛津燕尾式鞋面，将休闲和正装两种不同的风格创新的结合起来。
A woodblock heel grounds a tri color Oxford with air holes innovate mix with formal and casual.

运动鞋带

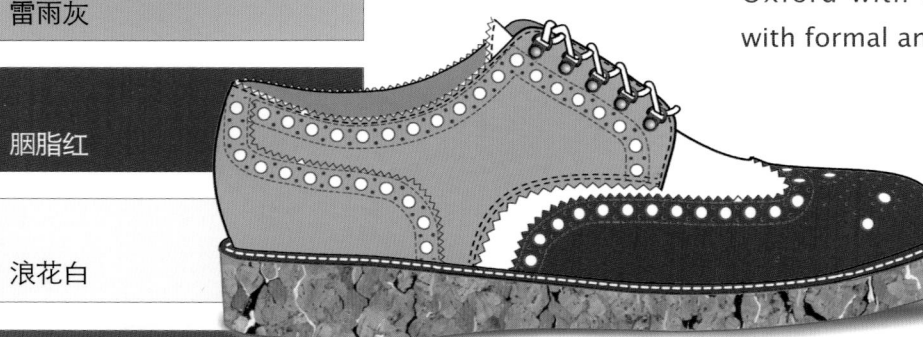

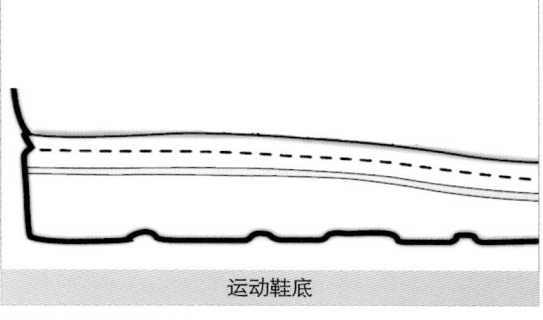

运动鞋底

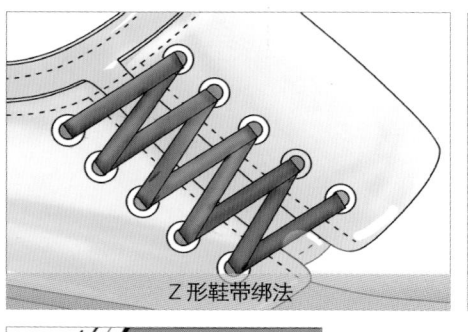

Z 形鞋带绑法

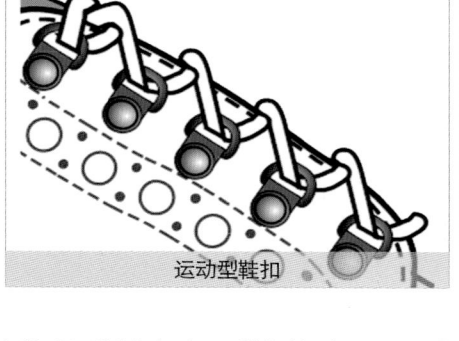

运动型鞋扣

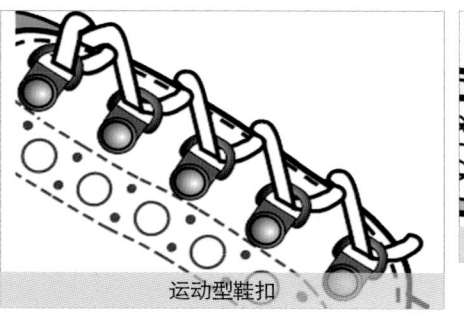

编织装饰

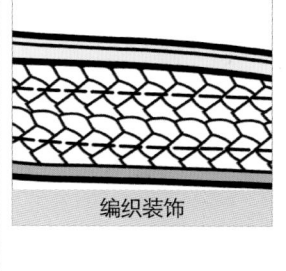

马克线缝法

包边处理

休闲鞋在软木面料上加上了彩色的透明 PVC, 鞋底也是软木材质。
Casual footwear with soft fabrics, a comfy contoured footbed and colored transparent PVC.

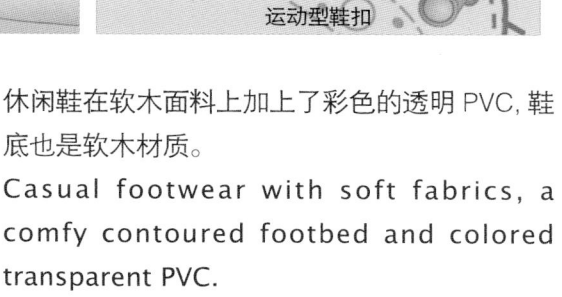

弹力松紧鞋口设计

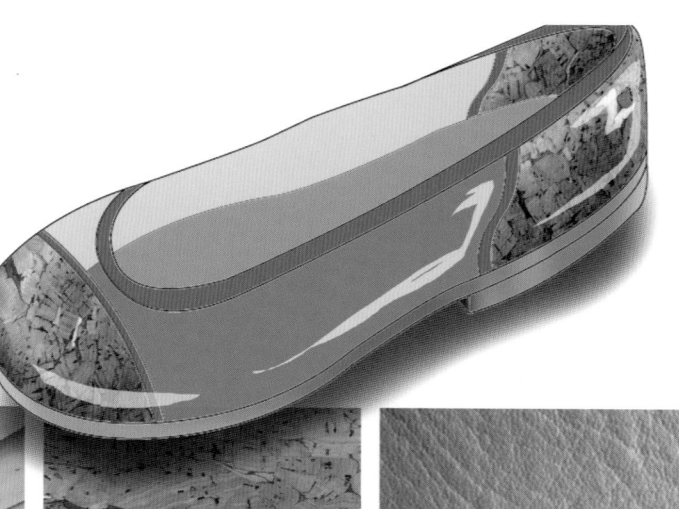

编织创新材料

手缝细线

印花木料

软木材料

牛皮

S 鞋履 2015 春夏海派时尚流行趋势

优活之梦
LIVING THE DREAM

优活之梦鞋履以舒适透气为主要出发点，配以纯手工，原生态的设计理念。低调不张扬却又独具魅力，从此让你爱上行走。平民贵族鞋履主要以精致的皮料搭配麻布等普通面料；生态材质则注重透气性及穿着的舒适性。编织作为主要手法在鞋履中也大放异彩，节俭设计的鞋履中提倡环保面料的再次利用。

Shoes in "LIVING THE DREAM" collection are made with breathable materials and simple design, low-key but charming. When you start to wear them, you're going to love walking. "ORDINARY NOBILITY" collection mainly focuses on using exquisite leather with linen and other common materials. "ECO MATERIAL" pays more attention to ventilation and comfort. We use weaving a lot this time. Eco-friendly materials are also used in "SIMPLE DESIGN" collection.

FOOTWEAR
2015 SPRING/SUMMER STYLE SHANGHAI FASHION TREND
海派鞋履流行趋势

2015 SPRING/SUMMER STYLE SHANGHAI FASHION TREND / **FOOTWEAR**

关键要素 | KEY POINTS

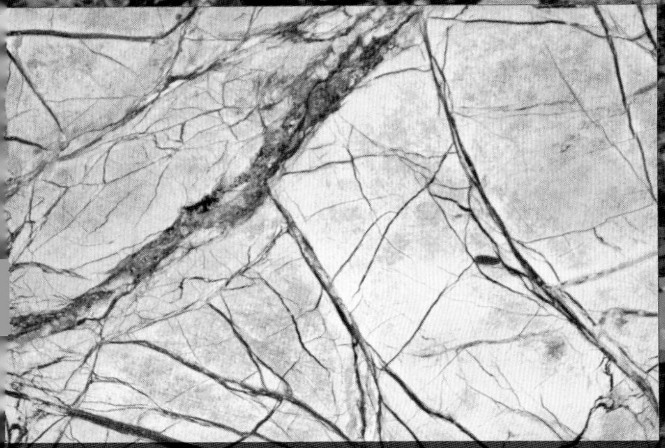

织补风尚
BACK TO THE BASICS

平民贵族
ORDINARY NOBILITY

生态材质
ECO MATERIAL

家庭手工
FAMILY HANDICRAFTS

淡奶黄

抹茶绿

苔藓绿

曜石黑

青花蓝

孔雀蓝

水晶蓝

豆浆白

香芋灰

紫砂棕

23

S 鞋履 2015 春夏海派时尚流行趋势

优活之梦
LIVING THE DREAM

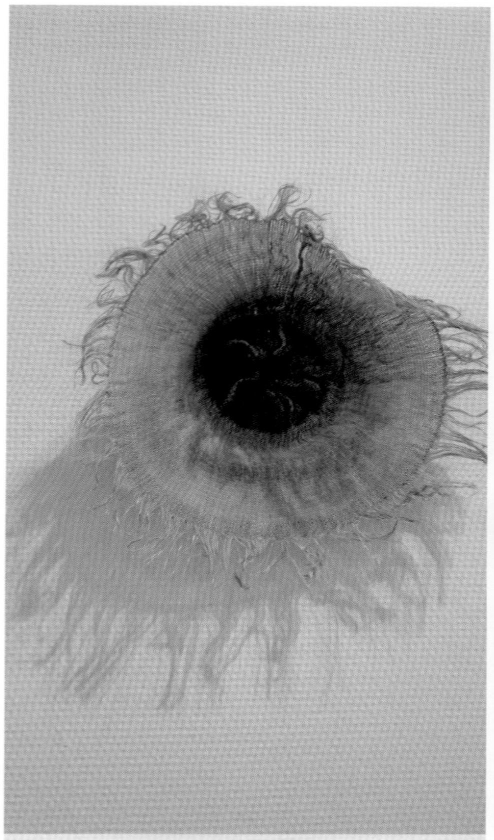

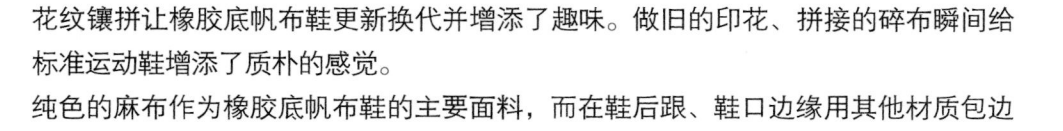

木质水果篮 　　　　瓦楞纸绘画 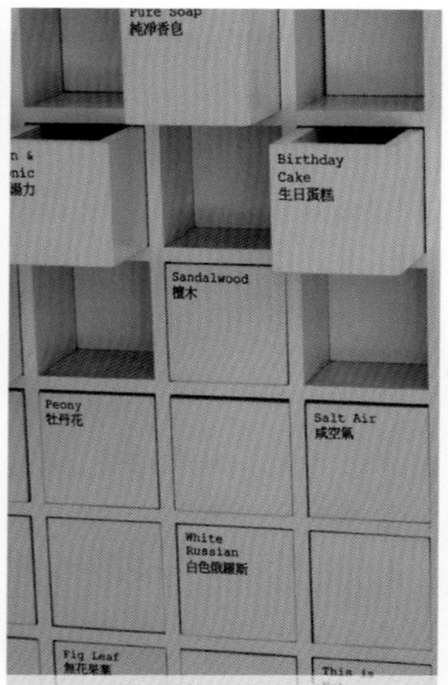　　　　简约分类收纳盒

银丝纤维墙面装饰品　　　　木质餐具 　　　　麻绳编织艺术品　　　　牛皮纸服装

设计灵感

花纹镶拼让橡胶底帆布鞋更新换代并增添了趣味。做旧的印花、拼接的碎布瞬间给标准运动鞋增添了质朴的感觉。

纯色的麻布作为橡胶底帆布鞋的主要面料，而在鞋后跟、鞋口边缘用其他材质包边点缀，精致又低调。

Washed fabrics and cloth patches put a fresh twist on a season-spanning sneaker with a feminine silhouette and a lightweight rubber sole.

Trimmed heel makes a sweet finish for a low-profile sneaker that will add sparkle to your steps.

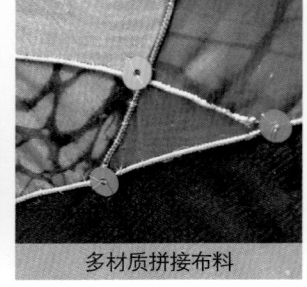

编织条带　　　　多材质拼接布料　　　　牛皮纸

24　www.style.sh.cn

2015 SPRING/SUMMER STYLE SHANGHAI FASHION TREND / **FOOTWEAR**

关 键 要 素 | 织补风尚
KEY POINTS | BACK TO THE BASICS

枣泥红

斗笠黄

剑麻灰

豆浆白

紫玉灰

鱼嘴女单鞋上采用类似红白蓝塑胶袋色系的宽条间色编织材质，采用柔和简约的亮色，让人联想起过往时代，彰显经典风格。

Wide stripes in bright and light red, white and blue radiate over the open toe of a graceful peep-toe pump.

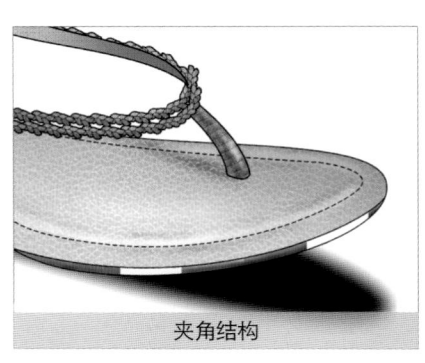
夹角结构

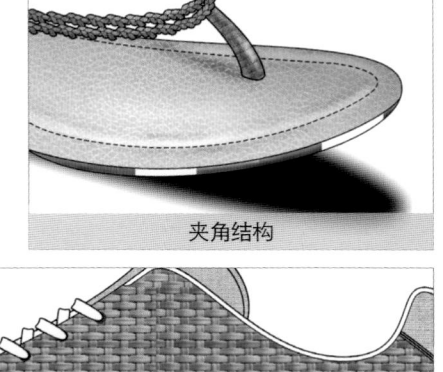
编织结构帮面

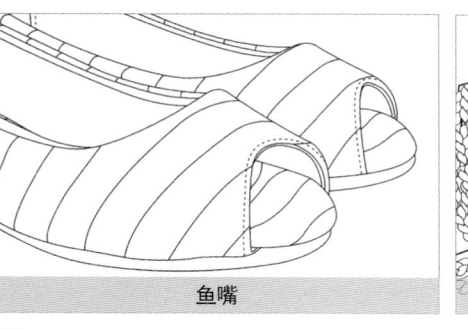
鱼嘴

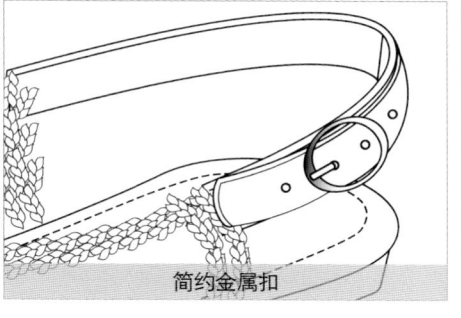
简约金属扣

改良豆豆鞋头

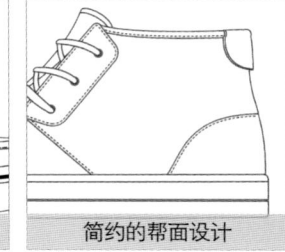
简约的帮面设计

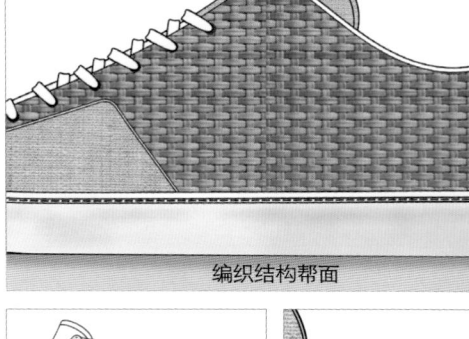
透气气跟

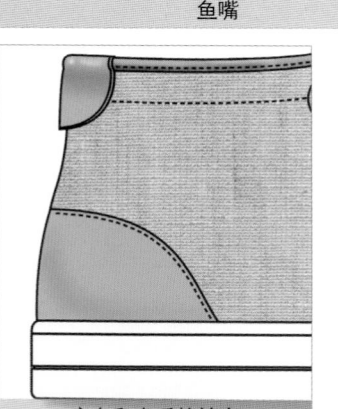
麻布和皮质的结合

简单的沙滩夹脚凉鞋，用最简洁的线条配上藤编的条带，仿佛置身于海滩边，简单的享受徐徐吹来的海风。

Braided straps sparkle atop a casual flip-flop. Wear it to a beach and enjoy delightful see wind.

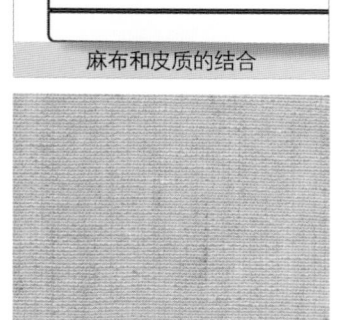
麻布

竹编材质

泡泡纱拼接

拼布

凉席编织材料

25

 鞋履 2015 春夏海派时尚流行趋势

优活之梦
LIVING THE DREAM

精致的复古灯具

繁复的大门装饰

木质的留声机

复古挂饰

独具匠心的工艺品

陈旧的手工艺品

华丽的灯饰

设计灵感

优雅尖头拖鞋是一种看上去很庄重，实际又比较随意的鞋款，也是平民贵族睡衣风貌最直接的代表鞋型，把高鞋面用于带有杯形鞋跟的拖鞋上，通过经典鞋底而让设计更为圆润，但鞋面上采用帆布、皮革或编织纤维面料暴露了其休闲自然的放松感。

A sound cup-shaped heel shapes a pointy-toe sandal cast either in canvas, leather or woven fabrics. Sometimes you can see women wear them together with pajamas in alleys, adding verve to casual wear.

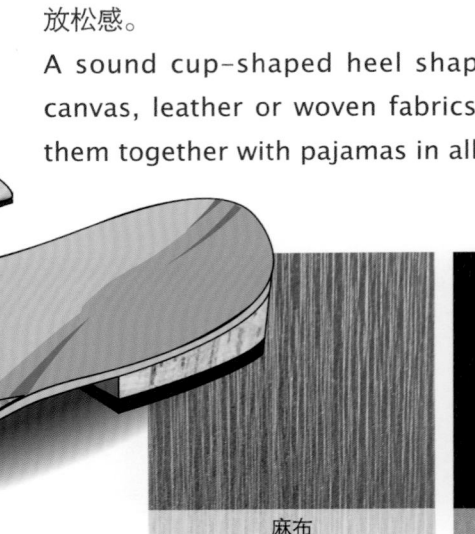

麻布

牛皮

缎面布

2015 SPRING/SUMMER STYLE SHANGHAI FASHION TREND / **FOOTWEAR**

关 键 要 素 | 平民贵族
KEY POINTS | ORDINARY NOBILITY

水晶蓝

青花蓝

抹茶绿

剑麻灰

香芋灰

在外耳式正装男鞋上用隐约的蓝灰色牛仔布印花图案搭配原始的木质鞋底，将严肃化为休闲，休闲中又带有严肃。
Bluish grey printed denim finishes a pair of dress shoes with a wood sole for day-to-night versatility.

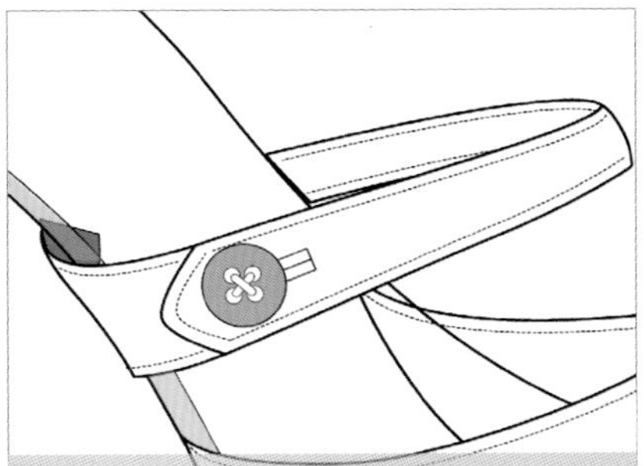

布艺纽扣

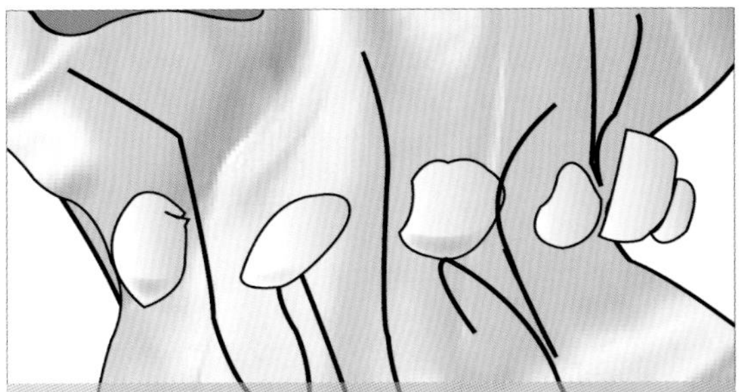

抽带结构

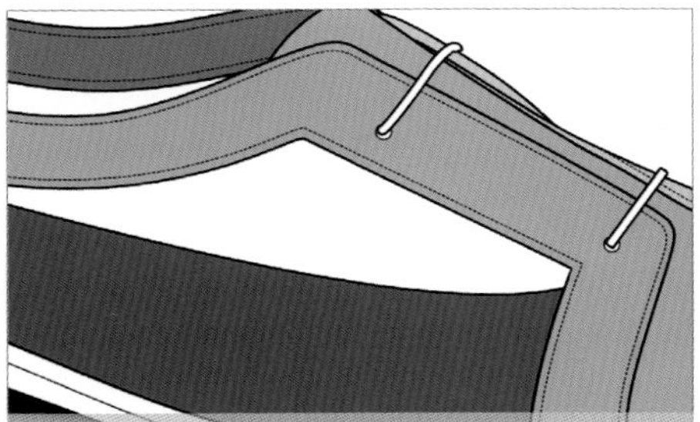

镂空侧面

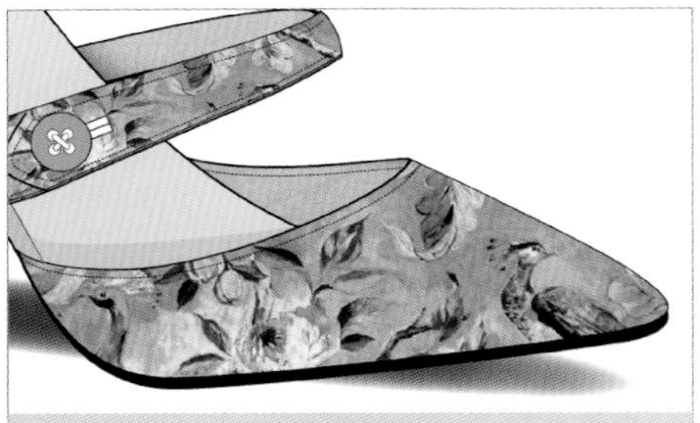

印花绸缎帮面

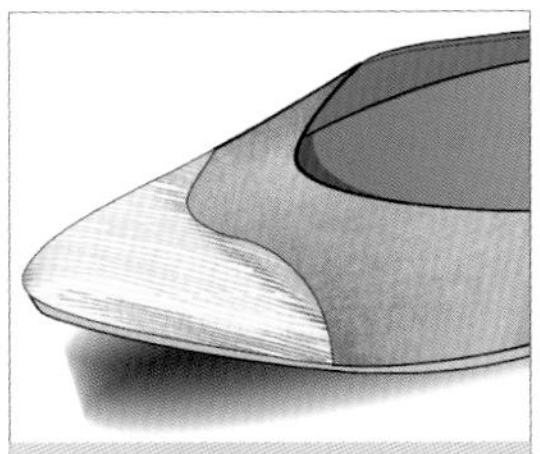

金属尖头鞋

绣花装饰

木鞋底

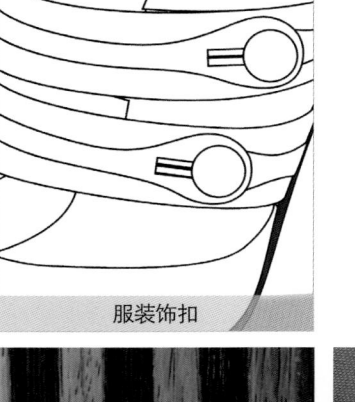

服装饰扣

造型简单的男士凉鞋采用硬化鞋底、鞋头略尖、创意的侧面镂空，内嵌式弹性鞋带也让穿脱更方便。
Hard outsole adds durability to a sandal ventilated for cool comfort and secured with inlayed shoestring.

间色木纹

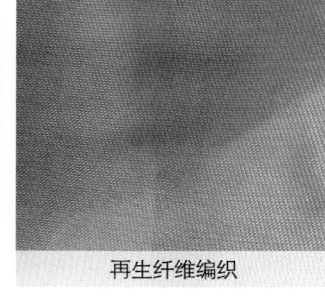

再生纤维编织

牛仔布拼花

印花褶皱布

珠光亮面布

S 鞋履 2015 春夏海派时尚流行趋势

优活之梦
LIVING THE DREAM

缝纫机和旧皮箱 　　旧上海贵妇雕塑 　　旧上海贵妇用品

淡雅的油画　　民族风格的陶瓷艺术 　　简约的床榻设计 　　电影场景模型

设计灵感

浅色麻灰让鞋装保持干净利落，搭配白色鞋带，打造质朴的外观。
马丁靴和牛津鞋款格外具有夏日休闲的气息 ，边缘处理上采用燕尾式花边，花型镂空增添了高端韵味。

Light grey with white shoestring makes shoes look quite clean and fresh.Side scrolls and engraving flower finish the iconic style of Martin Boot and Oxford.

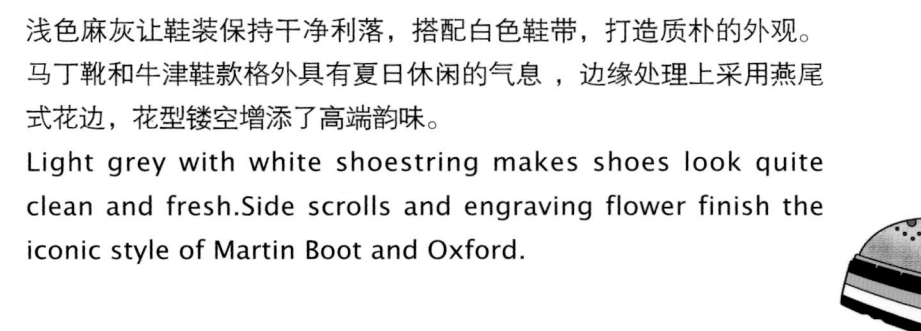

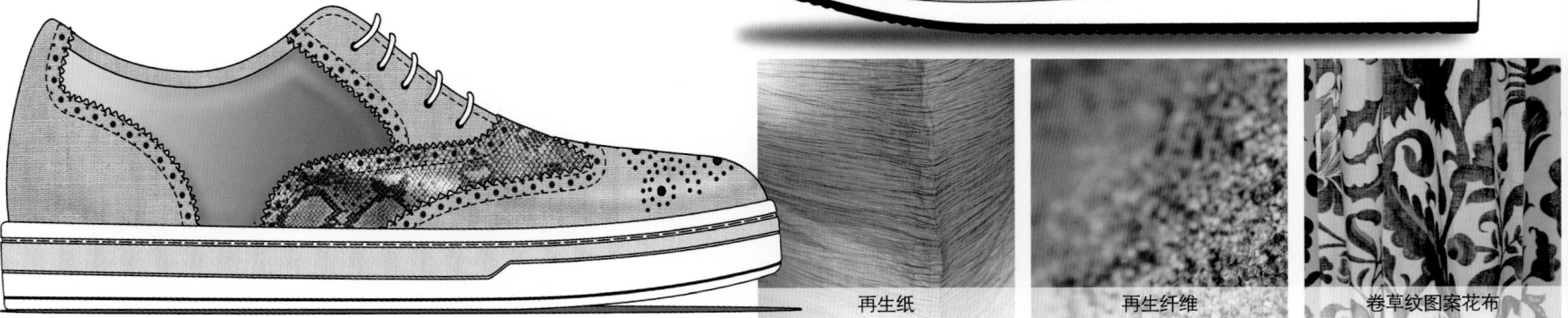

再生纸　　再生纤维　　卷草纹图案花布

28　www.style.sh.cn

2015 SPRING/SUMMER STYLE SHANGHAI FASHION TREND / FOOTWEAR

关键要素 | 平民贵族
KEY POINTS | ORDINARY NOBILITY

拼接的动物皮纹增添了精致的感觉，让质朴中多了一些贵气。

Animal skin texture adds verve to classic style.

蕾丝缎带带来活泼的、细腻的细节，随着春天的到来，干净的帆布质感透过蕾丝显现出来，低调而华丽。

Grosgrain lace band details this pair of canvas shoes comfortable for spring.

紫砂棕

斗笠黄

剑麻灰

豆浆白

香芋灰

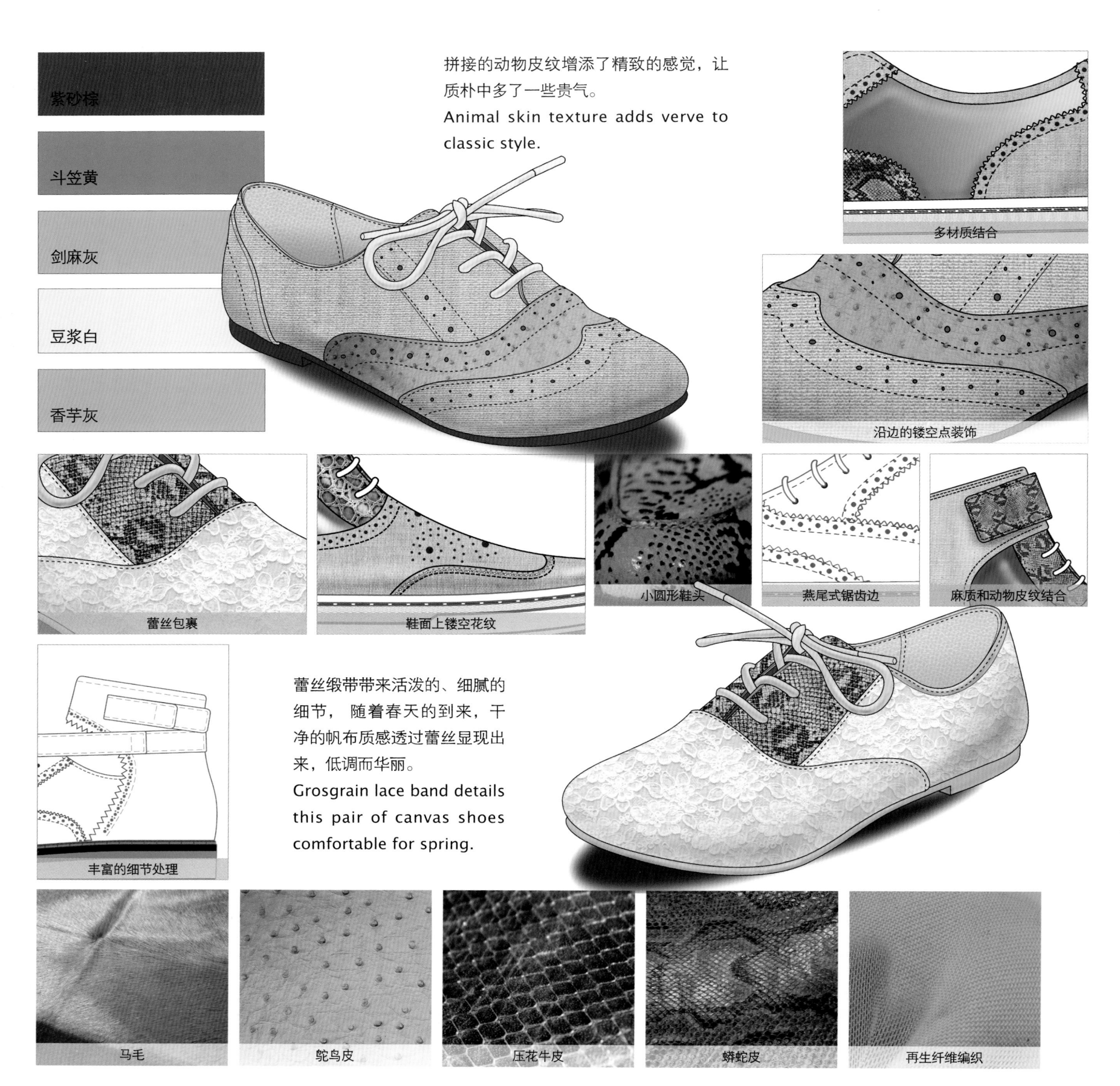

多材质结合

沿边的镂空点装饰

蕾丝包裹

鞋面上镂空花纹

小圆形鞋头

燕尾式锯齿边

麻质和动物皮纹结合

丰富的细节处理

马毛

鸵鸟皮

压花牛皮

蟒蛇皮

再生纤维编织

S 鞋履 2015 春夏海派时尚流行趋势

优活之梦
LIVING THE DREAM

仿自然的灯具设计

映日荷花

藤编的装饰球

石头墙面

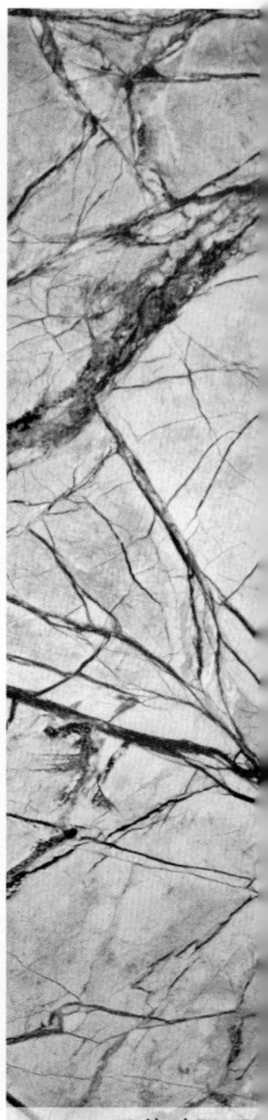

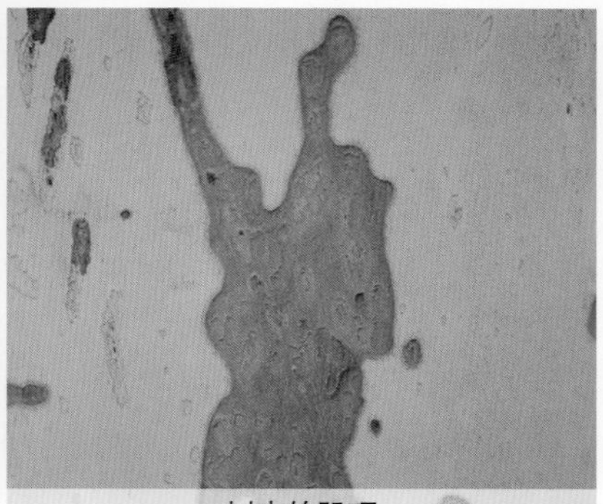

树皮的肌理

鹅卵石中生长的植物

干荷叶肌理

设计灵感

软木的结构使得软木制品有非常好的弹性、密封性、隔热性、隔音性、电绝缘性和耐摩擦性，加上无毒、无味、密度小、手感柔软、不易着火等优点，可用于外底、鞋床和鞋面，编织条带镶边为鞋面增添细节。

Cork leather has very good attributes, such as elasticity, sound and electric insulation and abrasion resistance. It's also non-toxic, tasteless, light, soft and non-flammable. It can be used in outsole, footbed and upper. Woven stripe trim details the vamp.

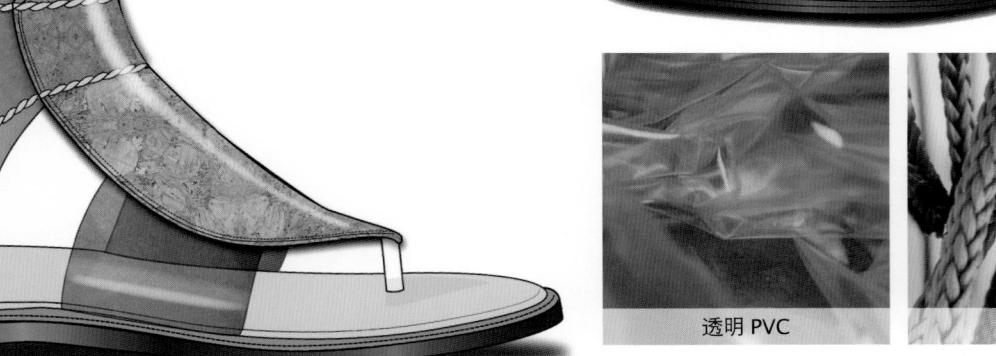

透明 PVC

藤编条带

特殊干花布料

30　www.style.sh.cn

2015 SPRING/SUMMER STYLE SHANGHAI FASHION TREND / **FOOTWEAR**

关键要素 ｜ 生态材质
KEY POINTS ｜ ECO MATERIAL

青釉灰

蜜瓜黄

剑麻灰

斗笠黄

紫砂棕

渔夫鞋款男女皆适宜，软木为鞋底夹层注入新颖的材料元素，全封口的鞋头和高鼻梁条带穿插结构显得非常自然。
Textured leather straps define a uni-sex fisherman-style sandal, crafted with a leather-wrapped latex foam footbed for comfort and support.

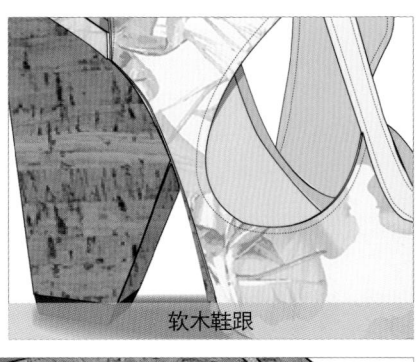
软木鞋跟

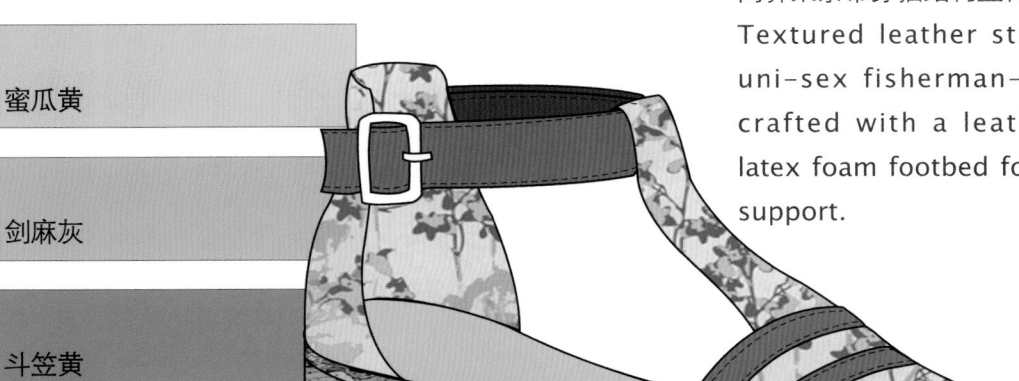

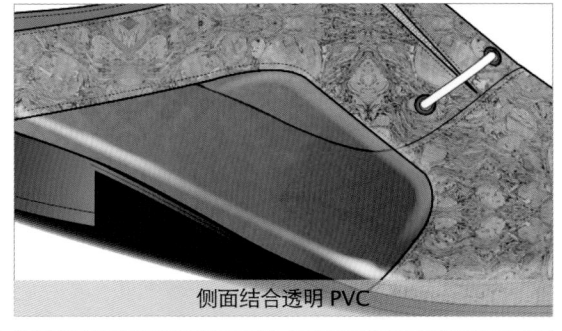
侧面结合透明 PVC

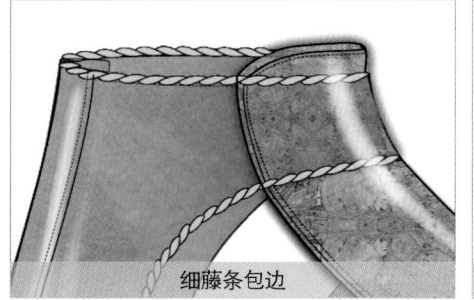
细藤条包边

透明 PVC 帮面

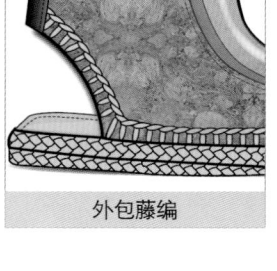
外包藤编

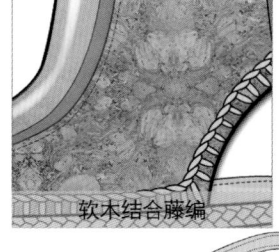
软木结合藤编

夹脚结构

软木大底

软木鞋底为单品注入休闲感，透气、舒适、柔软是自然最基本的诉求，鞋面采用清新淡雅的花纹，瞬间又有了文艺的感觉。
Beautiful flower prints add shine to a super-comfortable pair with cork sole anatomically engineered for optimal comfort and support.

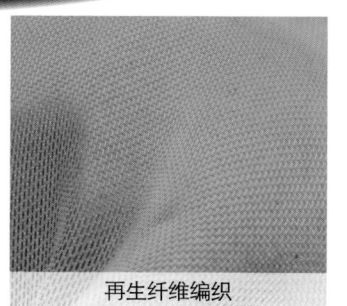

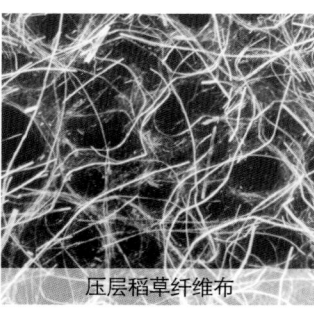
压层稻草纤维布

花卉图案大豆纤维布

软木材质

树脂木纹材质

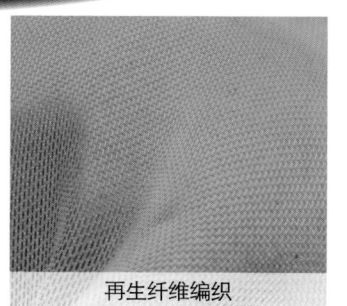
再生纤维编织

 上海设计之都
公共服务平台
认定专业平台

 海派时尚
STYLE SHANGHAI

鞋履 2015 春夏海派时尚流行趋势

优活之梦
LIVING THE DREAM

编织感的灯具设计 　条带图形面料座椅 　混排的马赛克墙砖

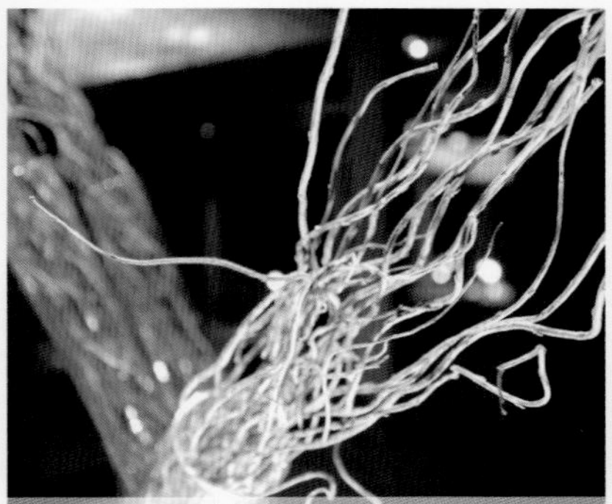

穿插结构的地砖　缠绕的藤条装饰 　手工灯罩 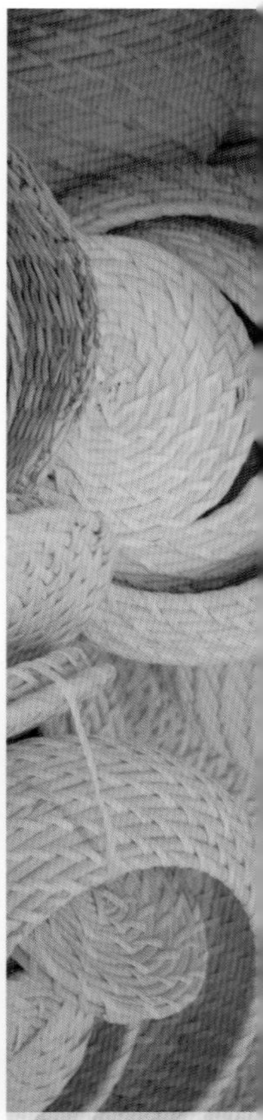　藤编工艺品

设计灵感

原生态的渔夫凉鞋，笼形渔夫款式覆盖面更大，梭织鞋面和细节为它们注入手工韵味，多重条带装饰的鞋面涵盖角斗士凉鞋和靴子构造。

Textured leather straps define a fisherman-style sandal, crafted with a foam footbed for comfort and support. The gladiator trend takes a curvy turn in a scintillating sandal with linked-leather architecture.

混色编织布　树脂编织材料 　激光切割的合成纤维布

2015 SPRING/SUMMER STYLE SHANGHAI FASHION TREND / **FOOTWEAR**

关 键 要 素 | 家 庭 手 工
KEY POINTS | FAMILY HANDICRAFTS

斗笠黄

蜜瓜黄

剑麻灰

孔雀蓝

青花蓝

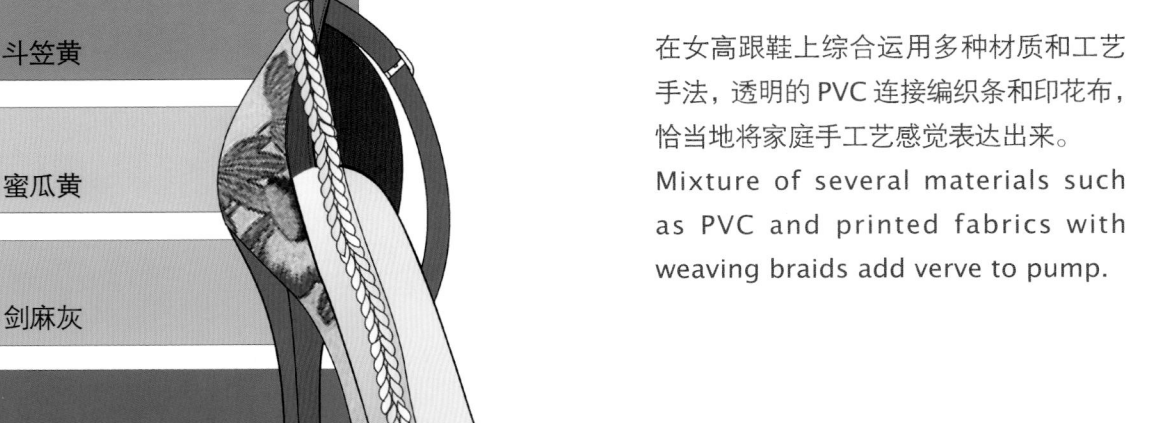

在女高跟鞋上综合运用多种材质和工艺手法，透明的PVC连接编织条和印花布，恰当地将家庭手工艺感觉表达出来。
Mixture of several materials such as PVC and printed fabrics with weaving braids add verve to pump.

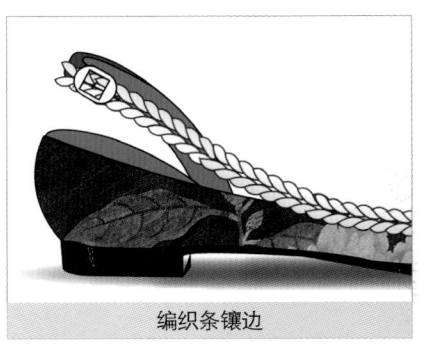
编织条镶边

编织条带结构

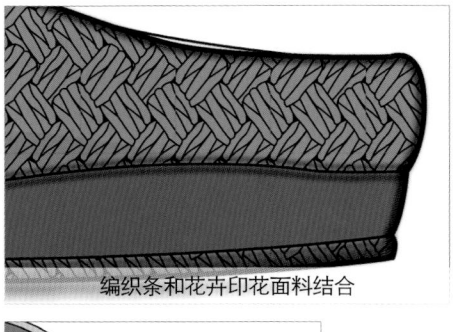
编织条和花卉印花面料结合

交叉穿插编织帮面

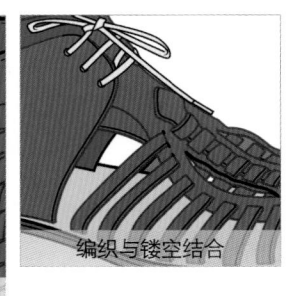
编织与镂空结合

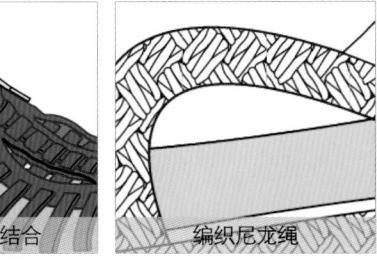
编织尼龙绳

透明 PVC 调和

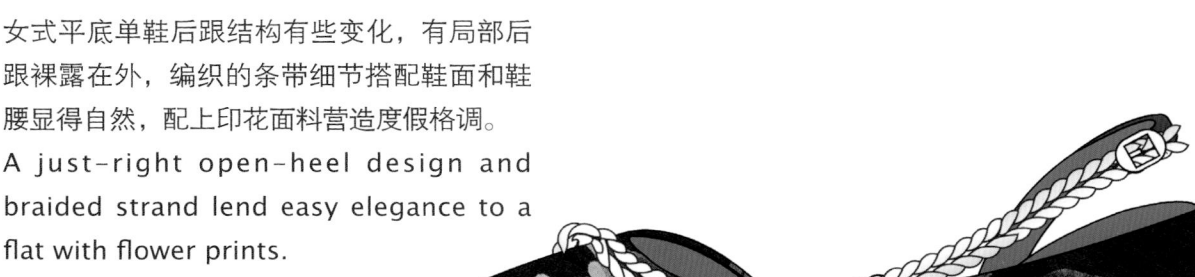

女式平底单鞋后跟结构有些变化，有局部后跟裸露在外，编织的条带细节搭配鞋面和鞋腰显得自然，配上印花面料营造度假格调。
A just-right open-heel design and braided strand lend easy elegance to a flat with flower prints.

鞋面皮料直接延伸到鞋跟

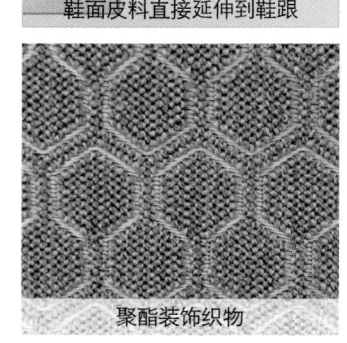
聚酯装饰织物

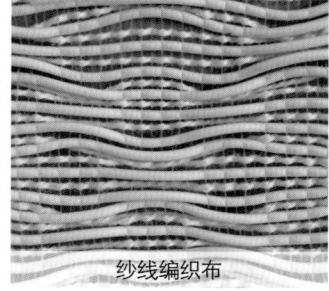
纱线编织布

聚酯藤编

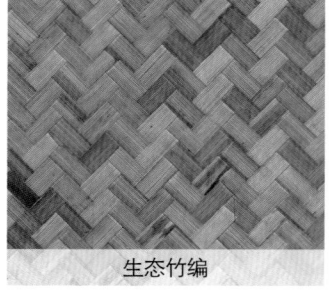
生态竹编

彩色藤编

鞋履 2015 春夏海派时尚流行趋势

时空筑梦
REALISING DREAMS THROUGH SPACE

坐落于浦西淮海中路最繁盛的商业区的环贸 IAPM 广场以她时尚前卫的室内装饰夺人眼球，营造出别具一格的超现实感的装潢艺术。犹如蜂窝般的仿生吊顶设计，更是让一进入这座巨型 shopping mall 的人眼前一亮。

IAPM mall is located on Huaihai Road in the busiest commercial district. Her stylish eye-catching interior creates a unique sense of surreal deco arts. People will also be attracted by the honeycomb-like ceiling designs as soon as they enter the mall.

FOOTWEAR

2015 SPRING/SUMMER STYLE SHANGHAI FASHION TREND
海派鞋履流行趋势

2015 SPRING/SUMMER STYLE SHANGHAI FASHION TREND / FOOTWEAR

关键要素 | KEY POINTS

复制粘贴
REPETITION & MODULIZATION

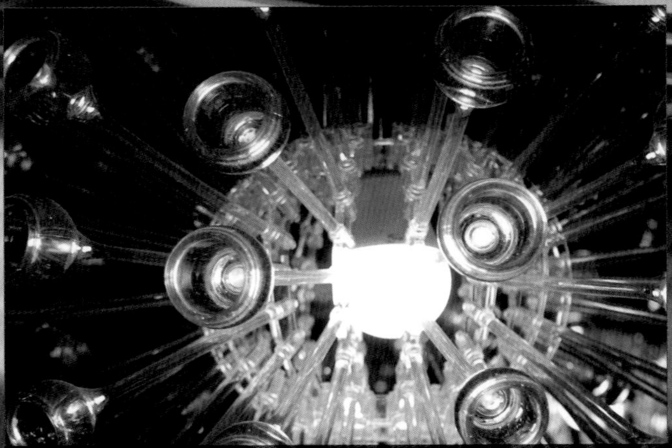

符号人生
SYMBOLIC LIFE

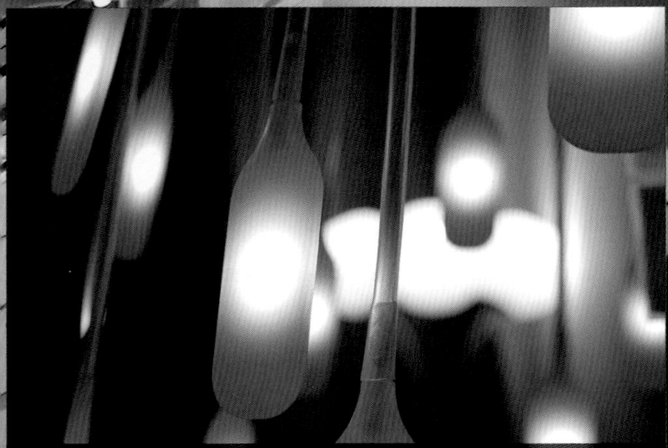

机械生物
MECHANICAL CREATURE

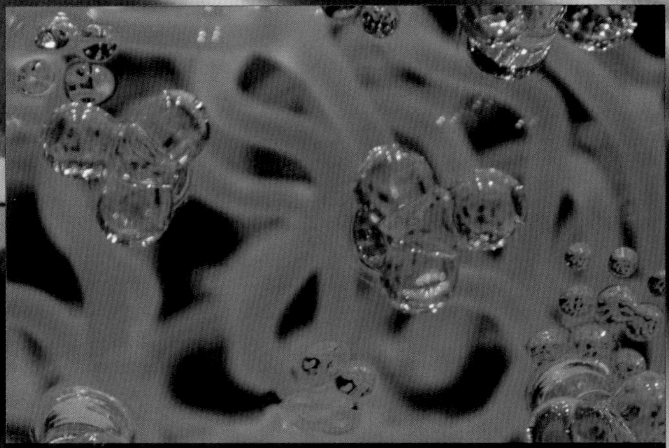

单纯设计
SIMPLE DESIGN

珍珠白

氯化灰

碳钢灰

天光蓝

光影蓝

亚铜灰

陨石棕

烟云灰

沙砾黄

柠檬黄

鞋履 2015 春夏海派时尚流行趋势

时空筑梦
REALISING DREAMS THROUGH SPACE

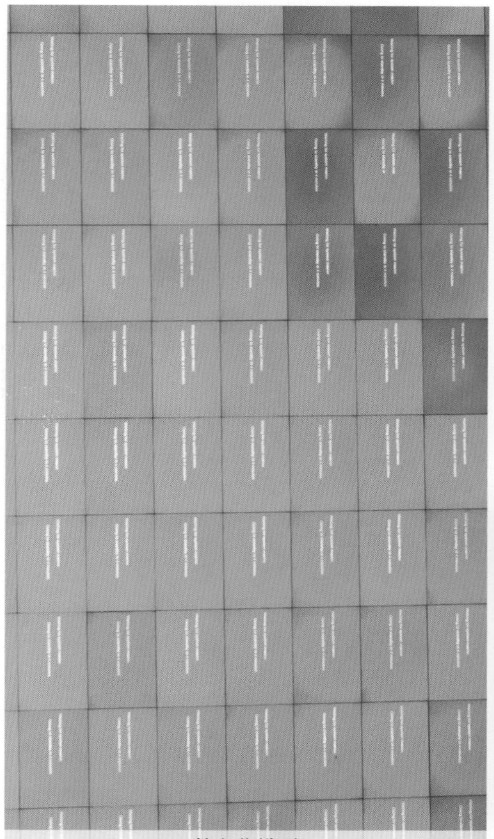

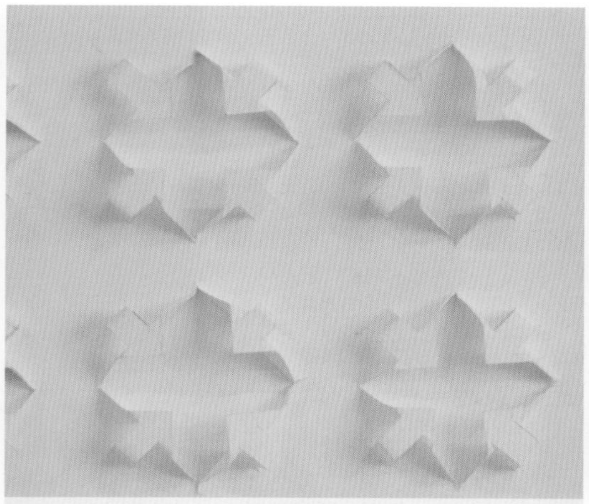

立构图案

旋转镂空

数位马赛克

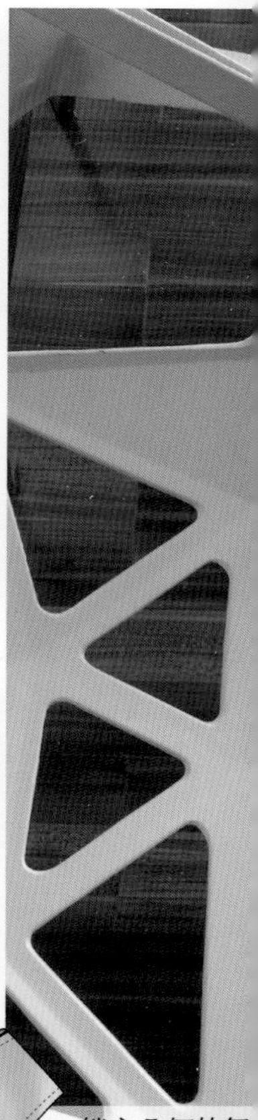

数字化排列

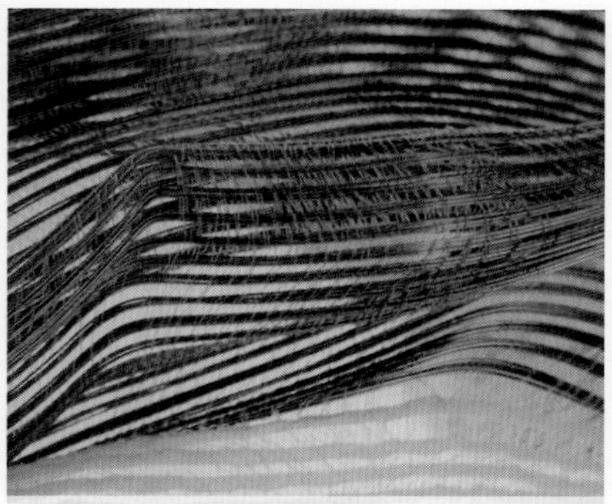

拉伸线条感

方圆有序排列

镂空几何的复

设计灵感

除了结构上的复制，在图形和颜色上也可进行设计交流。斜条纹的设计，简单的色块都符合未来人们对复制的设计构想。

Besides repetition of structures, grid overlays and colorful panels both can add visual texture for a retro-futuristic feel.

有色织带

键盘化

蜂窝状网布

36　www.style.sh.cn

2015 SPRING/SUMMER STYLE SHANGHAI FASHION TREND / **FOOTWEAR**

关键要素 | 复制粘贴
KEY POINTS | REPETITION & MODULIZATION

二维游戏图拼贴到鞋面上便形成了不同的视觉效果。只需轻微的斜度设计即可让简单的复制形成动感。
Shear mapping of elements for old video games brings new visual effects.

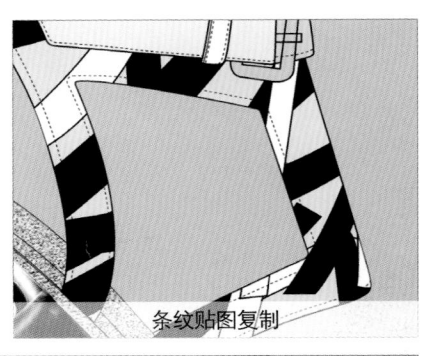

条纹贴图复制

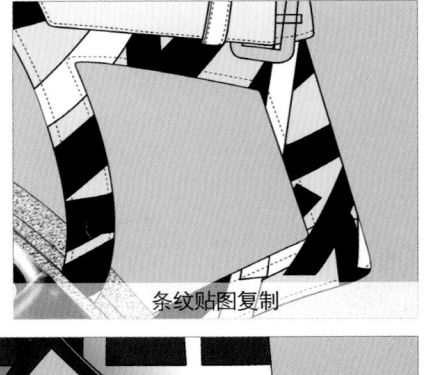

几何图复制

光影蓝

天光蓝

珍珠白

烟云灰

碳钢灰

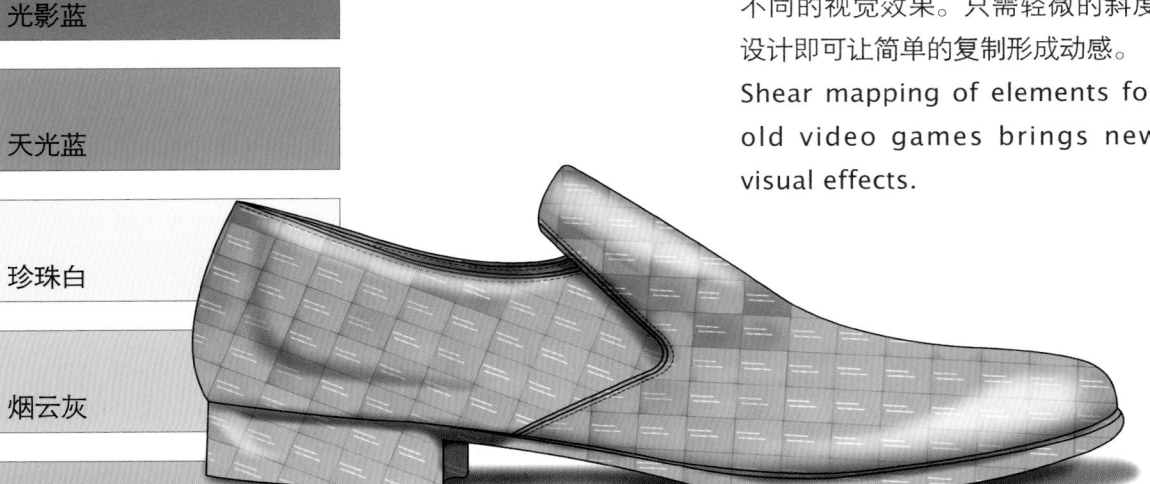

鞋跟立体复制

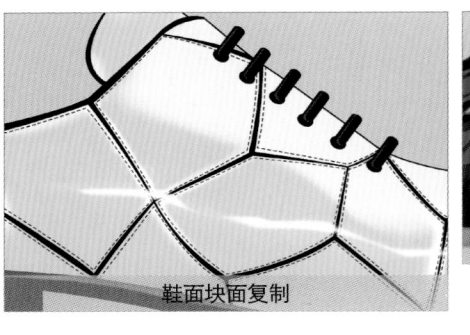

鞋面块面复制

帮面条带复制

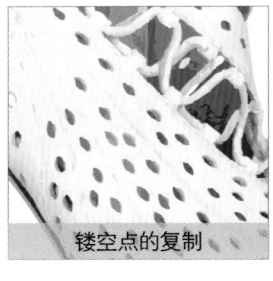

镂空点的复制

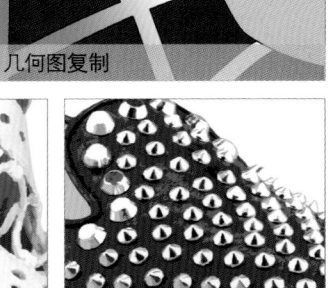

铆钉元素复制

蜂巢的六角形是这一季的设计灵感。在动物的规则图形上稍加修改，即可看到拼贴的复制感。
This season's collection is inspired by hexagonal honeycomb. Slight modification of nature graphics can bring collage sense.

鞋带构成的复制

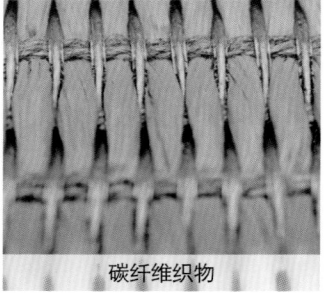

碳纤维织物

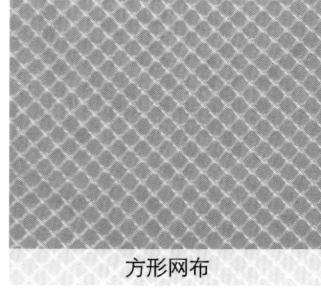

方形网布

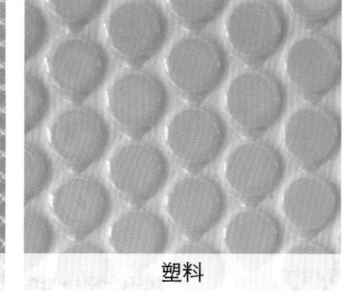

塑料

立体绣花

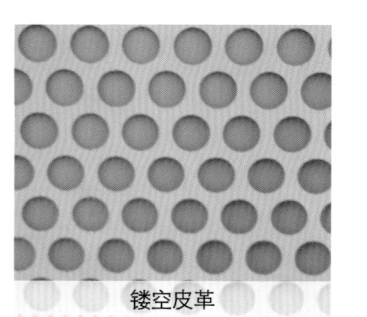

镂空皮革

上海设计之都
公共服务平台
认定专业平台

海派时尚
STYLE SHANGHAI

37

 鞋履 2015 春夏海派时尚流行趋势

时空筑梦
REALISING DREAMS THROUGH SPACE

组装复制

三维复制

视觉叠加

镜面复制

透叠复制

元素重复

光错觉

设计灵感

在鞋舌的设计上，运用多层叠加的方式即可达到未来复制感，同时以材料和贴图的方式强化表现，未来的童鞋是如此的 Fashion。
Multi-layer design adds retro-futuristic feel to shoe tongue emphasized by special material and collage.

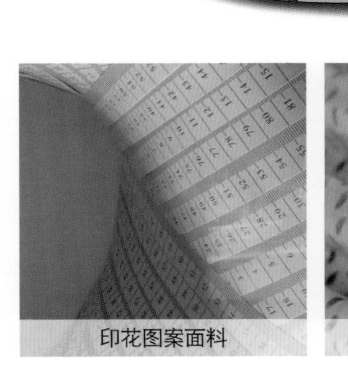

印花图案面料

镂空皮革

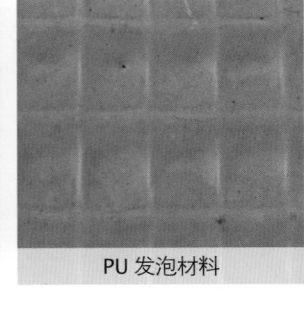

PU 发泡材料

38   www.style.sh.cn

2015 SPRING/SUMMER STYLE SHANGHAI FASHION TREND / FOOTWEAR

关 键 要 素 | 复 制 粘 贴
KEY POINTS | REPETITION & MODULIZATION

光影蓝

天光蓝

珍珠白

烟云灰

碳钢灰

多处层叠的运用是 2015 春夏的热点，不规则的拼接和网面的设计是运动系列的透气和时尚的选择。

The use of multi-layer is quite hot for 2015 Spring/Summer, irregular shaped panels and mesh design are choices for breathable sports fashion.

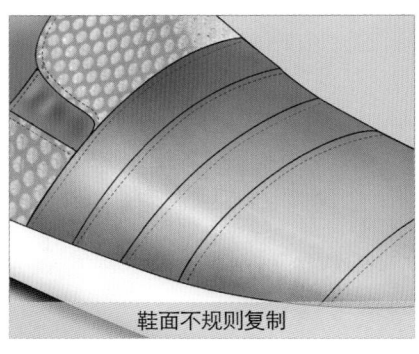
鞋面不规则复制

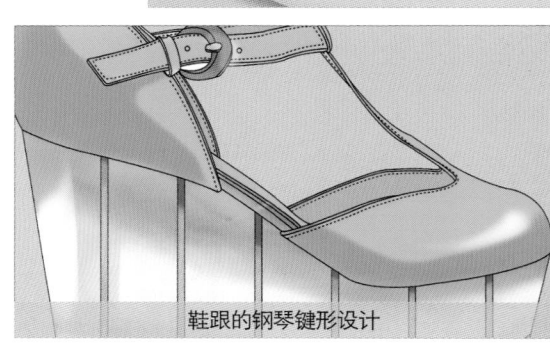
鞋跟的钢琴键形设计

鞋面方正图案的重复

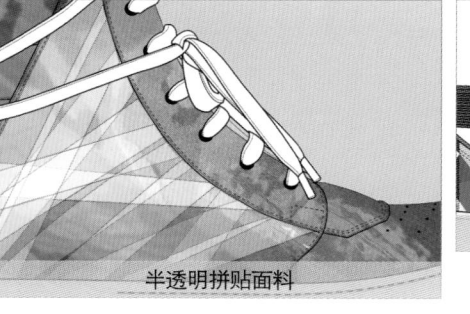
半透明拼贴面料

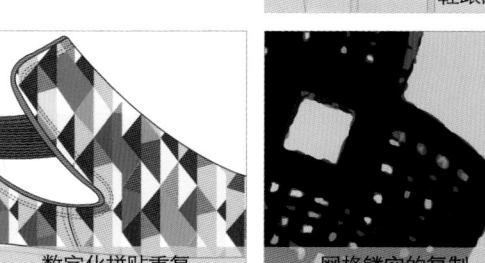
数字化拼贴重复

网格镂空的复制

鞋跟上的叠层设计

帮面装饰的重复排列

新一季的流行元素着重表现在帮口和鞋面的设计上，条带化图案成为主流，演绎时尚与人工的和谐美。

We focus on collar and vamp this season. Stripes are becoming popular.

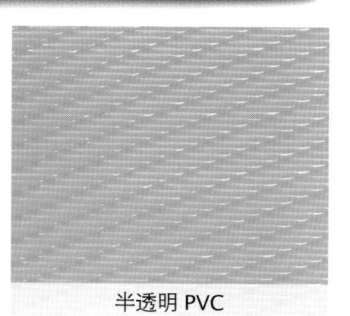

印花皮革做旧效果

帆布

编织尼龙

棉布

半透明 PVC

39

 **鞋履** 2015 春夏海派时尚流行趋势

时空筑梦
REALISING DREAMS THROUGH SPACE

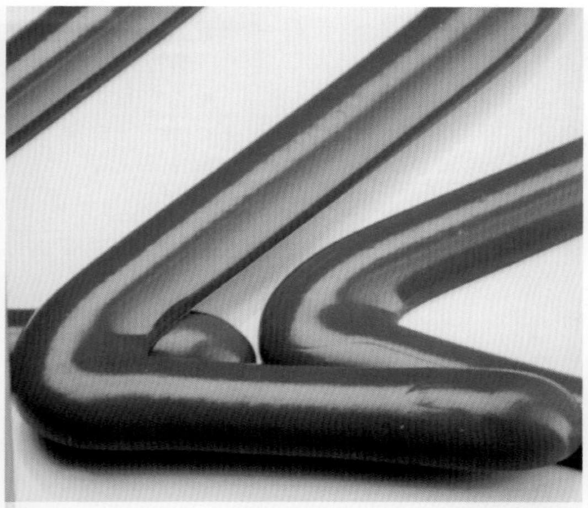

数字化符号

倒计时符号

环保循环符号

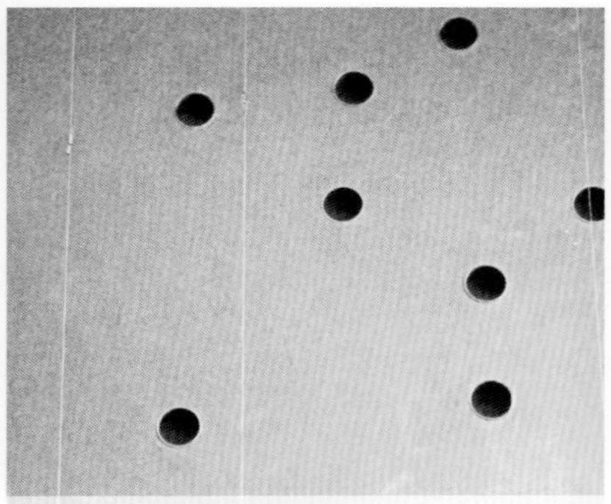

仿镜头视错觉

打洞感

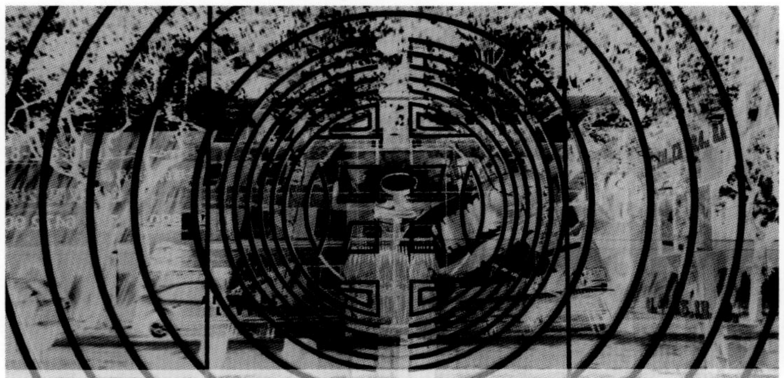

波纹数字化

块面等比

设计灵感

仿枪支设计，枪口式鞋跟，绑带式搭扣构成了符号的设计视觉效果，错觉符号感将是新一季的设计亮点。
Inspired by weapon, we innovatively design this heel. Strap-style buckle highlights visual effects.

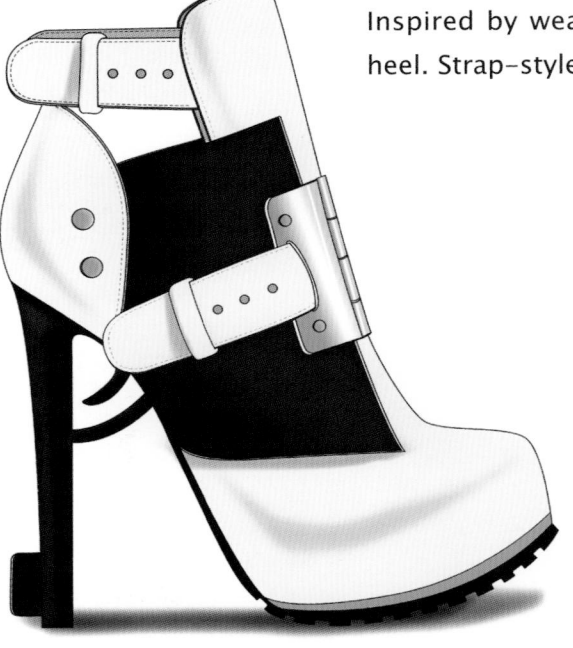

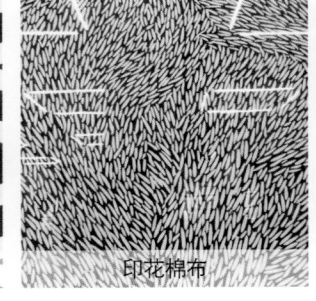
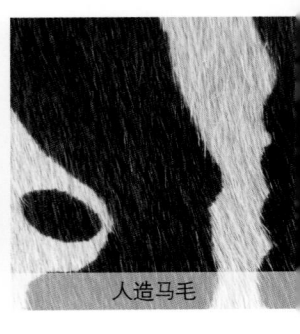

丝网印染布料

印花棉布

人造马毛

40   www.style.sh.cn

2015 SPRING/SUMMER STYLE SHANGHAI FASHION TREND / FOOTWEAR

关键要素 | 符号人生
KEY POINTS | SYMBOLIC LIFE

光影蓝

天光蓝

珍珠白

氯化灰

碳钢灰

电子元件！是的，用直接的电子元件图作为面料图案，配上简单的鞋帮设计即为符号的未来设计。
We surprisingly use electric components on vamp trying to convey retro-futuristic feeling.

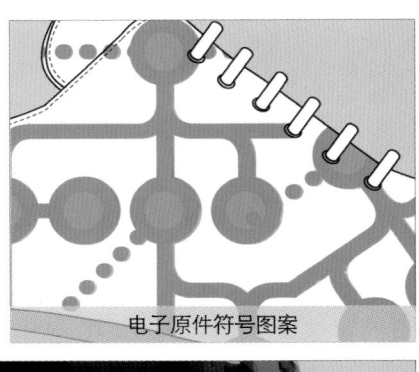

电子原件符号图案

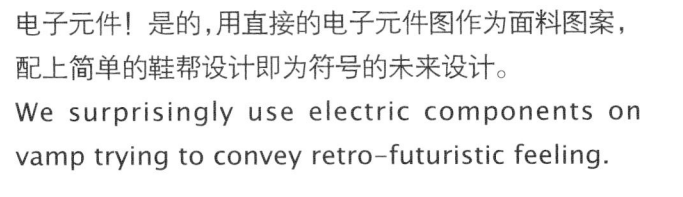

齿印在鞋底上的符号运用

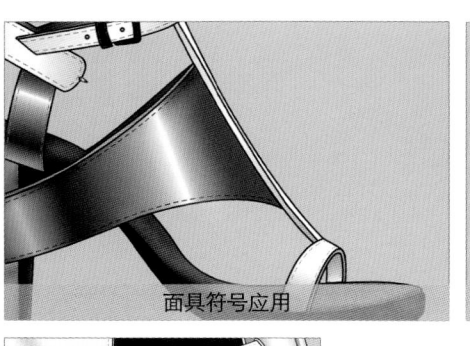

面具符号应用

星星在鞋跟上的运用

新动物皮

尖刺在鞋跟上的表现

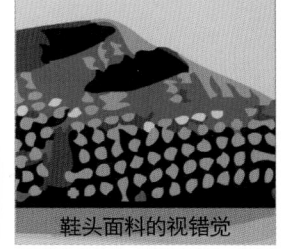

鞋头面料的视错觉

是的，眼睛！最新的设计表现在鞋面上会不会让你眼前一亮？只是简单的奶牛面料，因为做了效果后形成新的感觉就不一样了。
Black/white cow-print dyed vamp is quite eye-catching.

鞋跟的枪支形设计

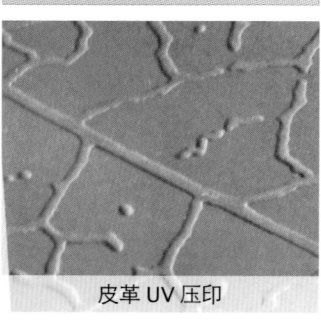

皮革 UV 压印

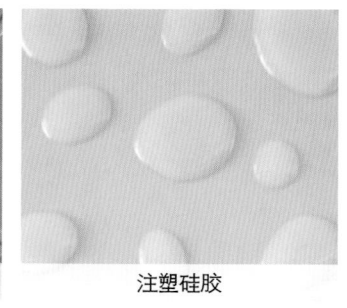

注塑硅胶

双色镂空面料

字母印花皮革

印花帆布

41

鞋履
2015 春夏海派时尚流行趋势

时空筑梦
REALISING DREAMS THROUGH SPACE

公共设计标示

楼道号

灯泡

立体方块

方圆组合

立体投影

人物

设计灵感

齿轮形鞋底在 2015 沙滩鞋的设计上拥有绝对的话语权。它既有防滑的功效，又兼具符号化的视觉语言。

A rubber sole bites into slick or wet terrain for four-wheel-drive performance from this cute and functional vehicle.

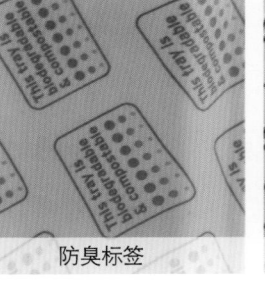

防臭标签

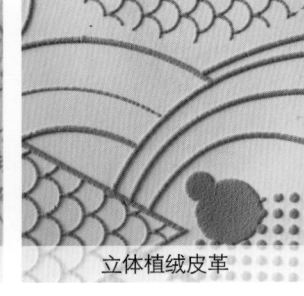

立体植绒皮革

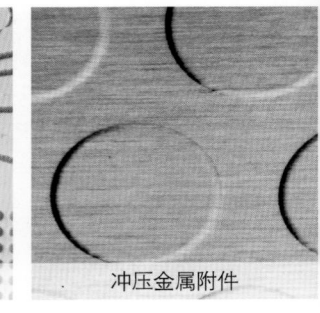

冲压金属附件

42   www.style.sh.cn

2015 SPRING/SUMMER STYLE SHANGHAI FASHION TREND / FOOTWEAR

关键要素 | 符号人生
KEY POINTS | SYMBOLIC LIFE

亚铜灰

天光蓝

日光白

氯化灰

碳钢灰

圆圆圈圈从小到大的排列、方圆的组合是元器件形的再设计，相对的，结构设计趋于简单以符合整体效果。
Different combination of various sizes of circles and squares add fabulous edge.

符号图形组合

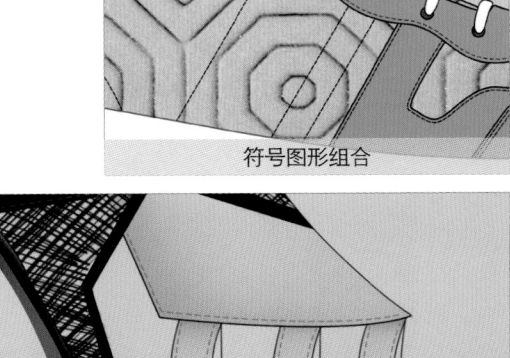
仿骷髅设计在鞋头上的应用

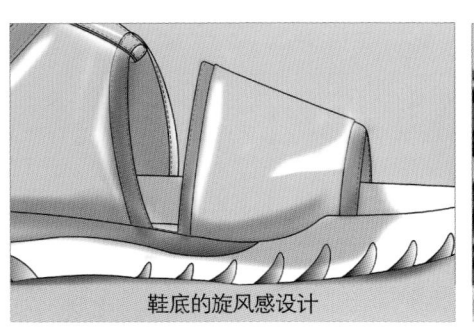
鞋底的旋风感设计

乌鸦尖嘴的符号化在鞋头上的表现

鞋跟上的面具设计

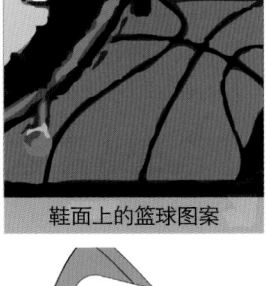
鞋面上的篮球图案

鞋跟上的箭头符号

符号化的表现突出点在于颜色的直接运用和图案的二维设计。将平面化的图案直接贴在鞋面上即有电子元件的高科技时代感。
Graphic design highlights symbolization and high-tech feeling.

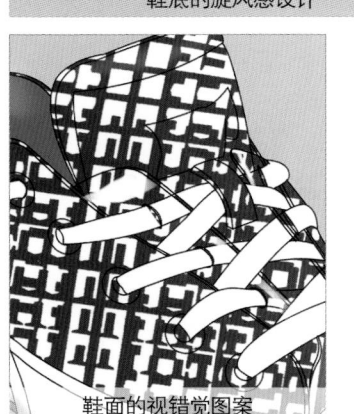
鞋面的视错觉图案

充棉车花布料

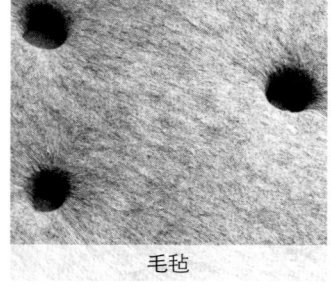
毛毡

黏胶鞋底

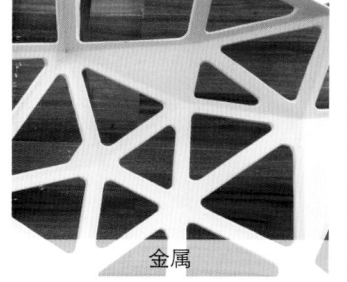
金属

塑料

43

鞋履 2015 春夏海派时尚流行趋势

时空筑梦
REALISING DREAMS THROUGH SPACE

钢筋外露

强化结构感

反射多色彩

冲刺激流感

吸盘机械感

仿太空屋顶

蛇纹外墙

设计灵感

绚烂的颜色交织透明的 PVC，用生态镂空的结构营造生物啃食的效果。仿机械的块面、尖锐的跟尖是机械的另一种表现形式，直接而炫目。

An aggressive sole adds some bite to colorful transparent PVC vamp.

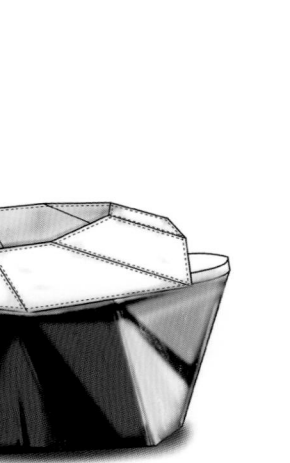

印花面料

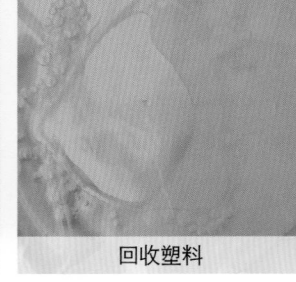

回收塑料

反光材料

44　www.style.sh.cn

2015 SPRING/SUMMER STYLE SHANGHAI FASHION TREND / FOOTWEAR

关键要素
KEY POINTS │ 机械生物
MECHANICAL CREATURE

陨石棕

亚铜灰

天光蓝

柠檬黄

光影蓝

2015 的男鞋偶尔也会有新鲜的设计。女鞋的高度和铆钉元素的运用直接表现在鞋头和后跟，从第一眼到最后一眼都能吸引人们的眼球。
Sometimes we also put bold design in men's shoes. For this one, a pump sole grounds a versatile wingtip touched up with rivet details to catch the eye.

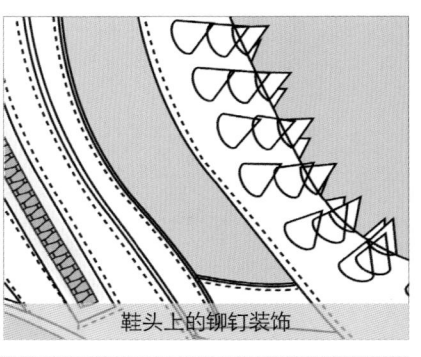
鞋头上的铆钉装饰

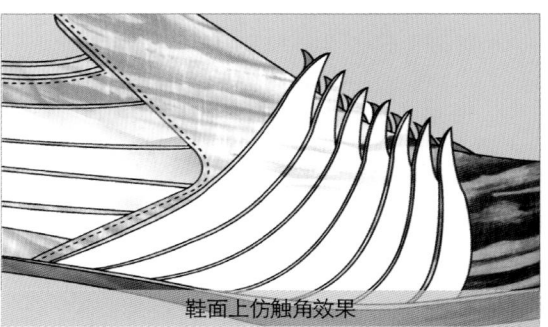
鞋面上仿触角效果

鞋面上的仿水细节

机械块面在鞋跟上的表现

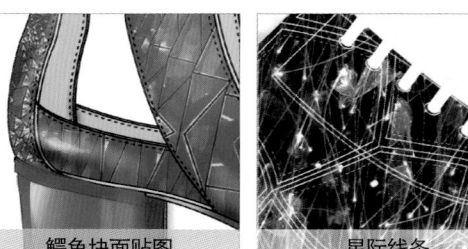
鳄鱼块面贴图

星际线条

宇宙虚幻的贴面

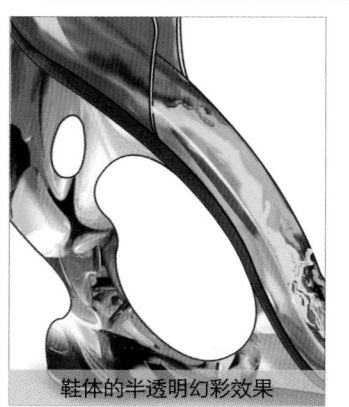
鞋体的半透明幻彩效果

传统的帮面做底，鞋面用片面结构装饰，形如章鱼的触角在你脚背游走，惊悚而刺激。
Give you an exotic adventure in this pair patterned with scales.

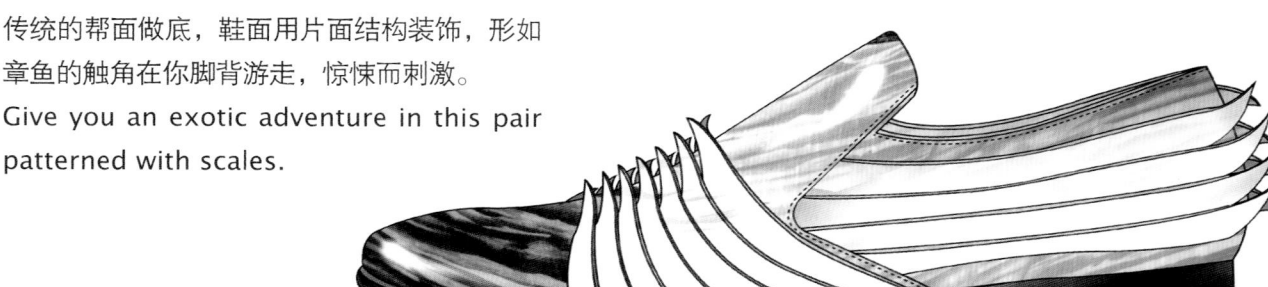

仿贝壳效果材料

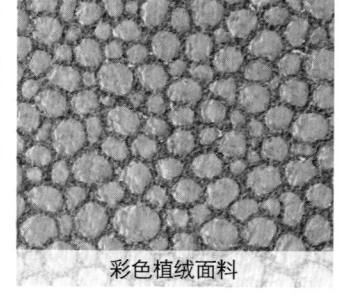
彩色植绒面料

手缝珠片面料

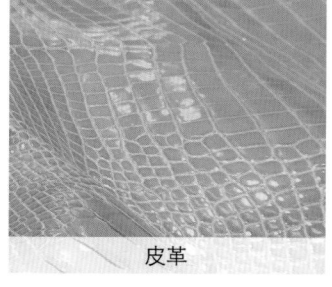
皮革

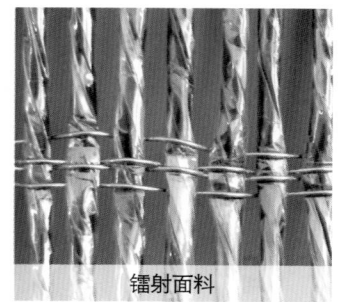
镭射面料

鞋履 **2015 春夏海派时尚流行趋势**

时空筑梦
REALISING DREAMS THROUGH SPACE

大块面拼接

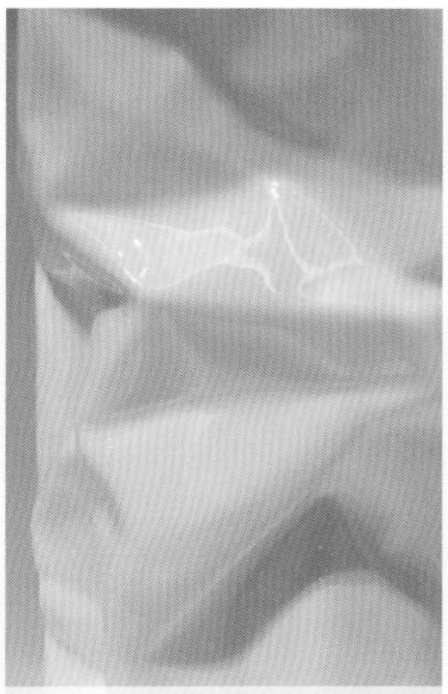

仿随意褶皱

一体化设计

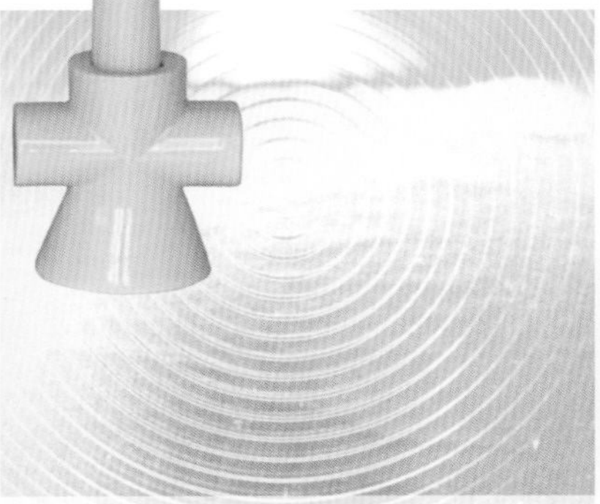

透明肌理

涡形单线循环

空间寂寥

白墙

设计灵感

单纯设计就是不需要过多的细节。大块包面，简单搭扣，直角跟型即满足了简单设计的原则，面料以细腻质感为主。

Less is more. Large surface, neat buckle and right heel angle with delicate fabric fashion a simple design.

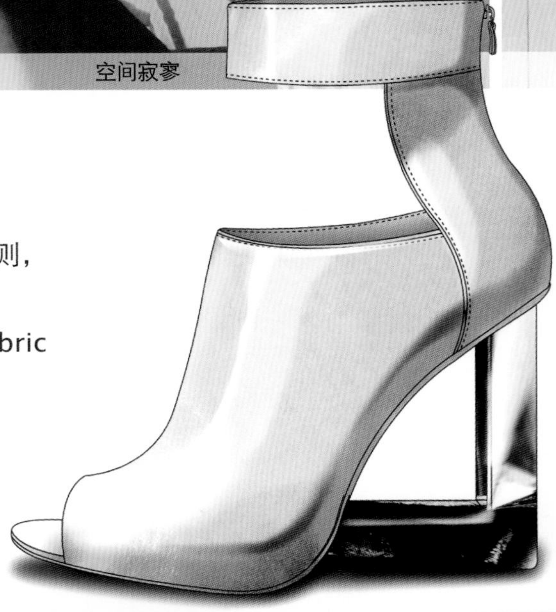

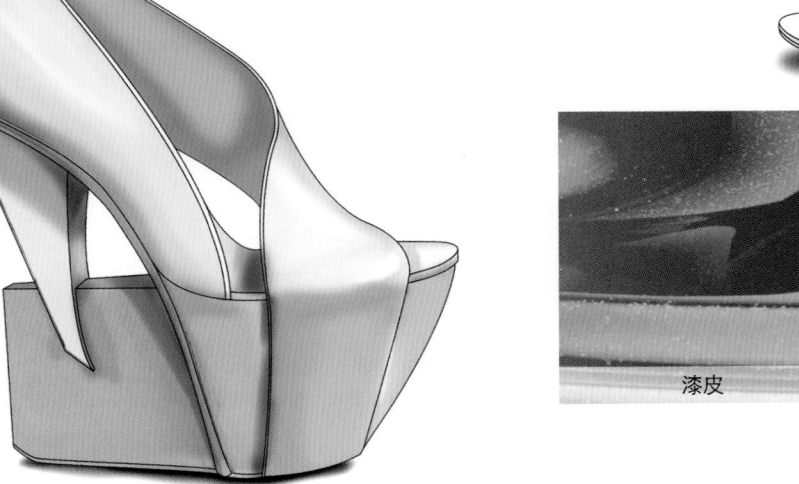

漆皮

斜纹面料

雾面金葱面料

46　www.style.sh.cn

2015 SPRING/SUMMER STYLE SHANGHAI FASHION TREND / FOOTWEAR

关键要素 | 单纯设计
KEY POINTS | SIMPLE DESIGN

烟云灰

氯化灰

珍珠白

日光白

碳钢灰

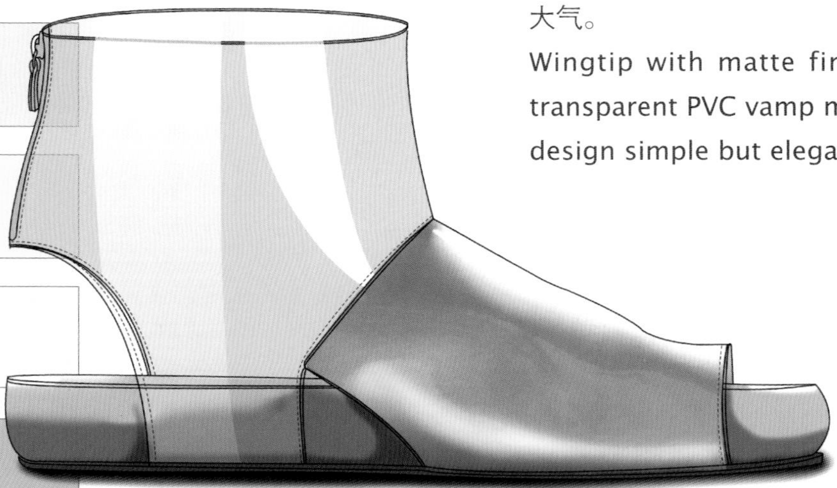

前后部简单设计，鞋头亮面哑光设计，鞋帮透明 PVC 设计，整体简单大气。
Wingtip with matte finish and transparent PVC vamp make this design simple but elegant.

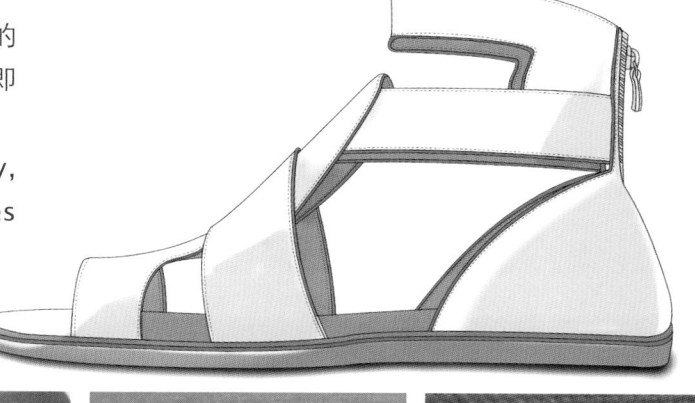

鞋面透明材质 PVC

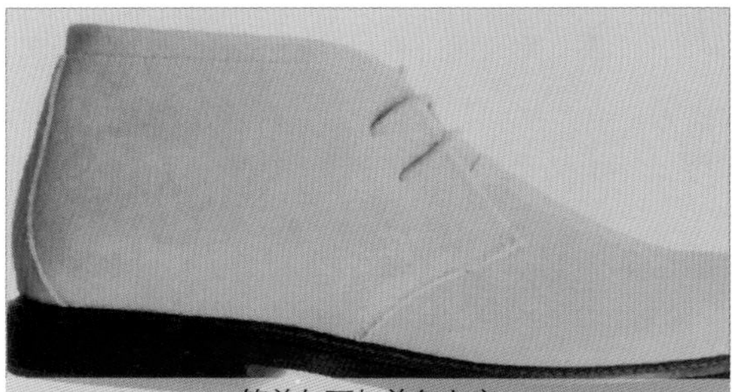

简单包面与单色大底

交替错落拼接

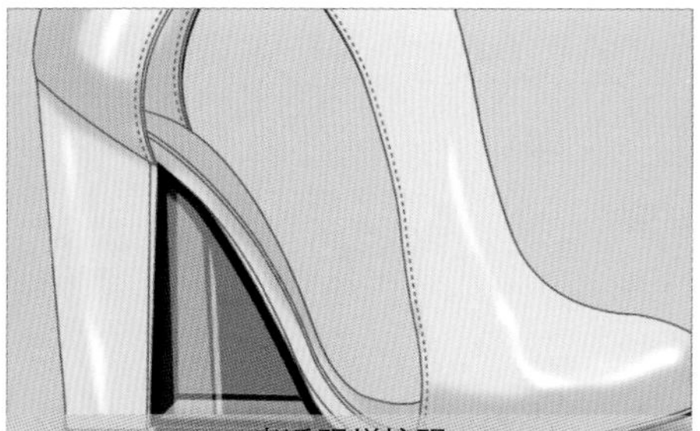

半透明拼接跟

大块面包跟

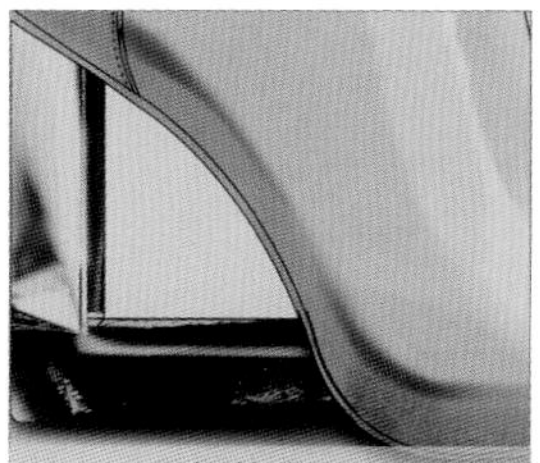

几何体金属跟

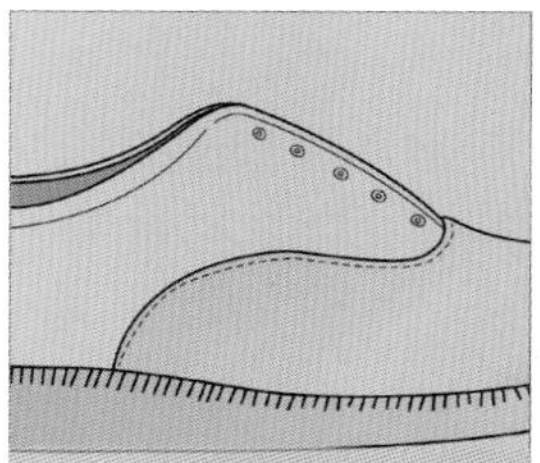

流畅线条

镂空跟

今年的单纯设计除了简单的结构，稍微规则的穿插是另一个设计点，纯白面料、极细描边即可表现鞋子的设计感。
Clean construction, elegant interplay, white tone or delicate silhouette defines simplicity in this year's trend.

压花皮革

纯白皮革

PVC 反光材料

棉布

牛津布

  **鞋履** 2015 春夏海派时尚流行趋势

时空筑梦
REALISING DREAMS THROUGH SPACE

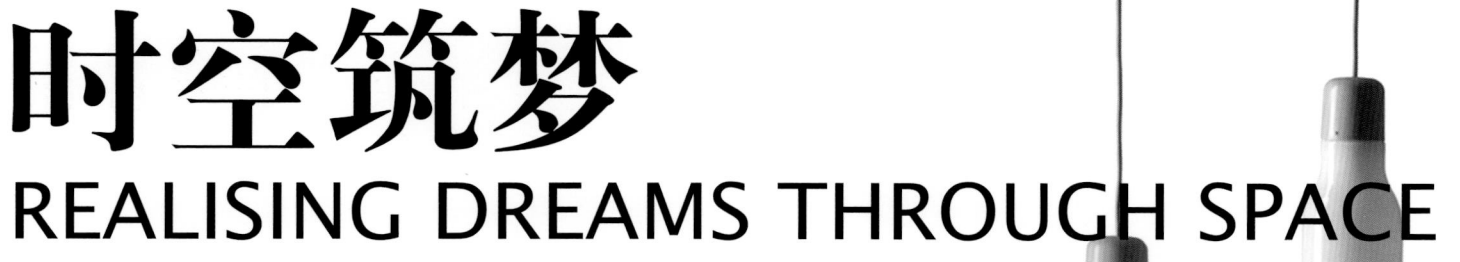

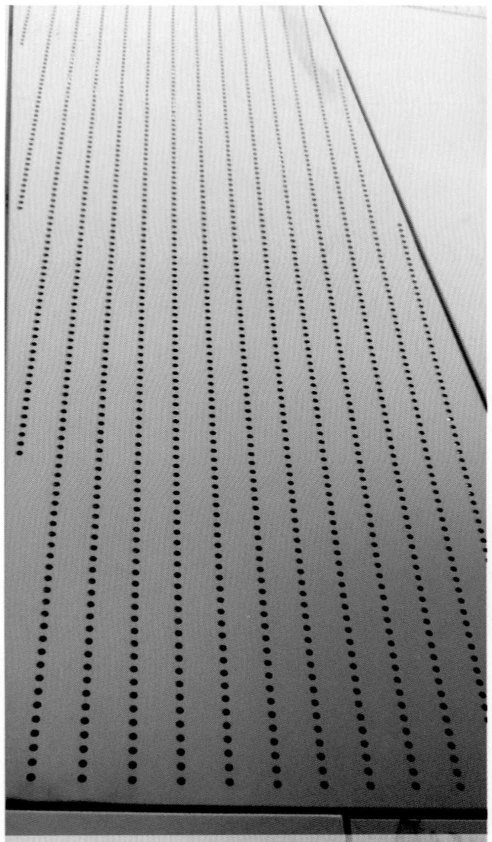

高楼外立面　　水溶性包装泡沫　　白色外壳 　　玻璃管 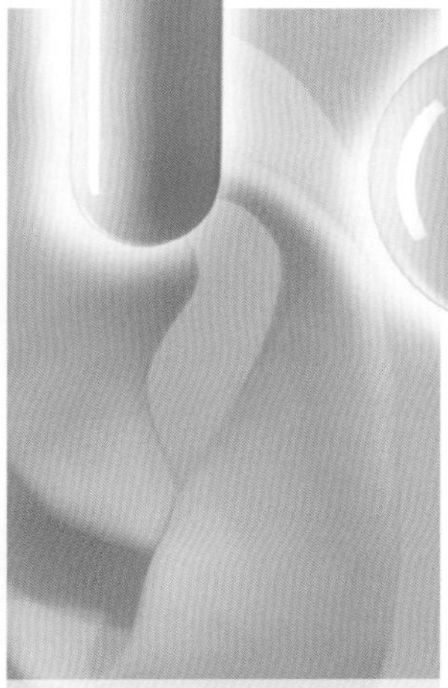　　雕塑　　简单地面设计　　平静的海

设计灵感

简单大块面，单色鞋底，包面设计构成了单纯男鞋的设计。面料上采用细颗粒感的未来材质来表现。

Single color sole, large surface vamp and rich grainy skin are fashioned into this design.

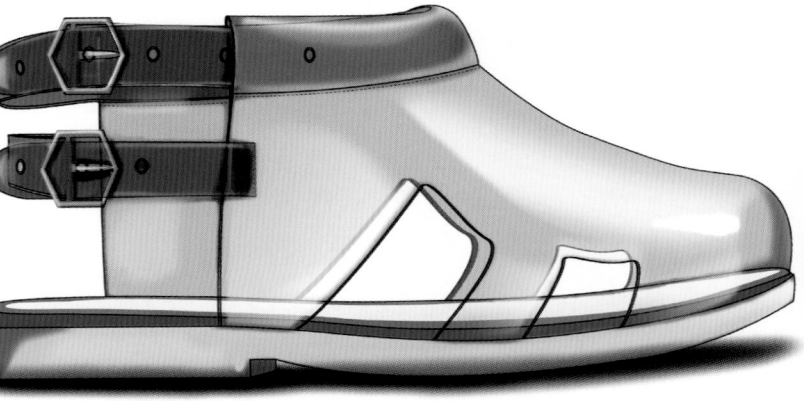

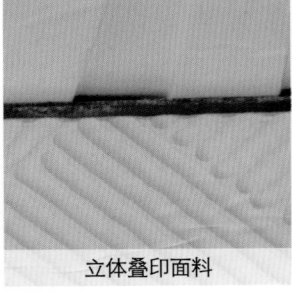
立体叠印面料

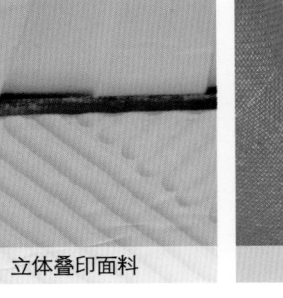
细纹路牛津布

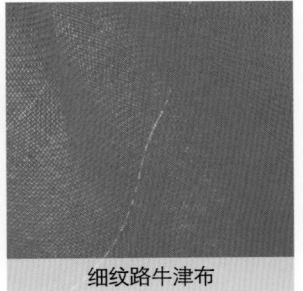
透气网面材料

48　www.style.sh.cn

2015 SPRING/SUMMER STYLE SHANGHAI FASHION TREND / FOOTWEAR

关键要素 | 单纯设计
KEY POINTS | SIMPLE DESIGN

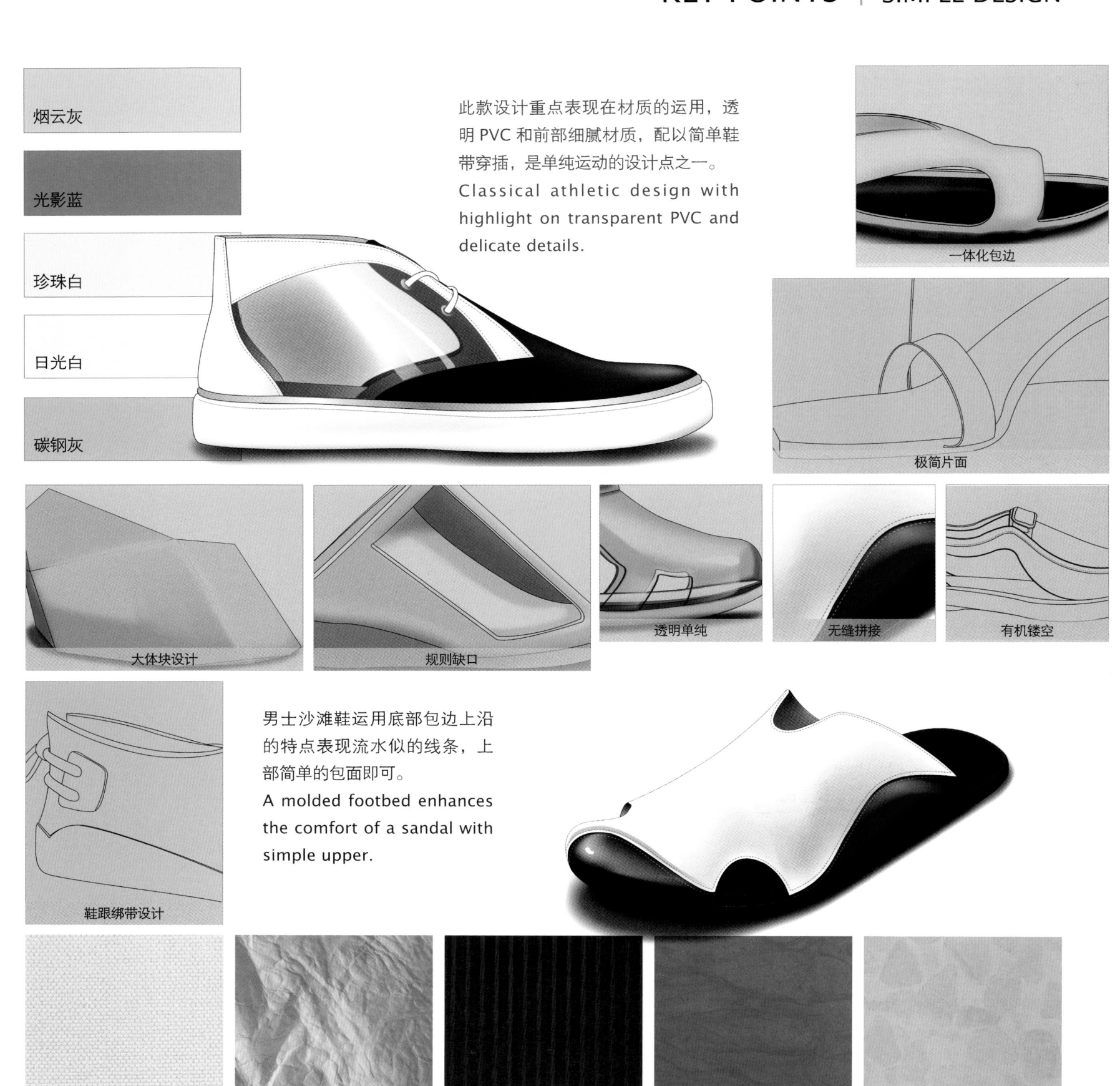

烟云灰

光影蓝

珍珠白

日光白

碳钢灰

此款设计重点表现在材质的运用，透明 PVC 和前部细腻材质，配以简单鞋带穿插，是单纯运动的设计点之一。
Classical athletic design with highlight on transparent PVC and delicate details.

一体化包边

极简片面

大体块设计

规则缺口

透明单纯

无缝拼接

有机镂空

男士沙滩鞋运用底部包边上沿的特点表现流水似的线条，上部简单的包面即可。
A molded footbed enhances the comfort of a sandal with simple upper.

鞋跟绑带设计

细孔梭织布

皱面羊皮

暗色条纹布

超细纤维仿皮革

暗纹提花布

多彩享梦
ENJOY COLORFUL DREAMS

由厂房、油漆字、工业零件等符号提炼出不同形态的 M50 三个字符，紧靠苏州河的地理位置和仓库式的存在形式，体现了 M50 是历史、文化、艺术、时尚与创意的融合。 这里更是吸引了大批具有鲜明个性特点的年轻人来此自由创作。

Here we symbolize M50, the name for a contemporary art district in Shanghai with elements from plant, paint characters and industrial parts. It is located close to Suzhou River and a warehouse–style area where history, culture, arts, fashion and creativeness integrate. A large number of young artists with distinct personalities choose to move their studios there.

FOOTWEAR

2015 SPRING/SUMMER STYLE SHANGHAI FASHION TREND

海派鞋履流行趋势

2015 SPRING/SUMMER STYLE SHANGHAI FASHION TREND / **FOOTWEAR**

关键要素 | KEY POINTS

童年拾趣
CHILDHOOD REFLECTIONS

跨界实验
TRANSBOUNDARY EXPERIMENTS

快乐崇拜
JOLLY IDOLATRY

都市街头
CITY STREETS

浅裸粉

蜜桃粉

幻彩红

冰山蓝

玛瑙蓝

松石绿

珊瑚红

迎春黄

琵琶橙

石榴红

51

鞋履 2015 春夏海派时尚流行趋势

多彩享梦
ENJOY COLORFUL DREAMS

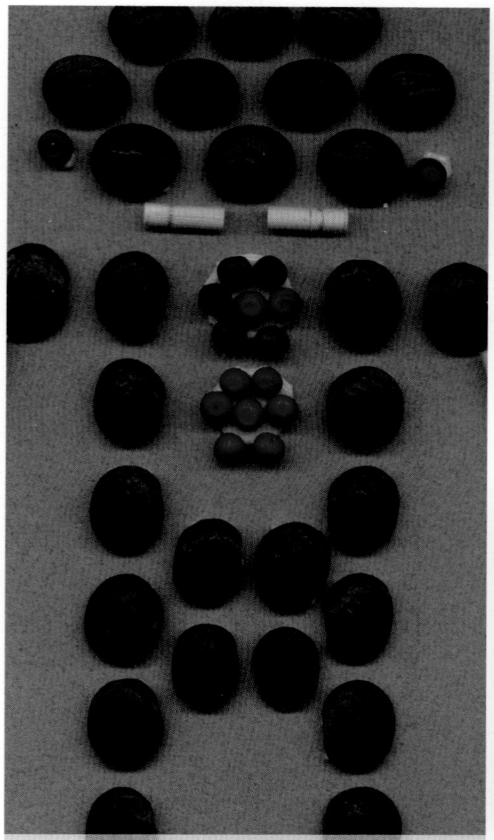

趣味性的摆放 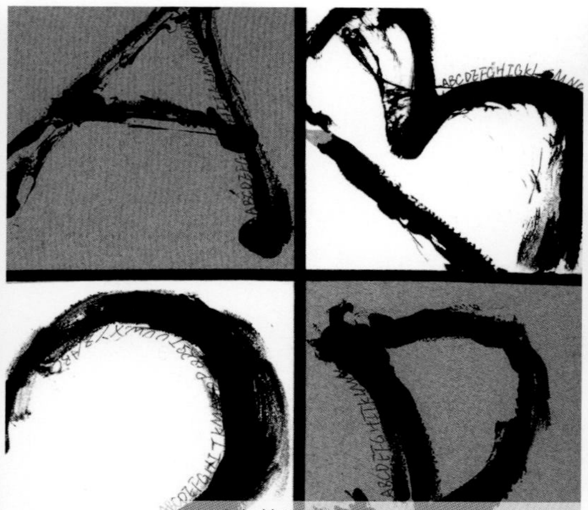　　　　　随意的手写字体 　　　　乐高的随意拼接 　　　儿童玩具

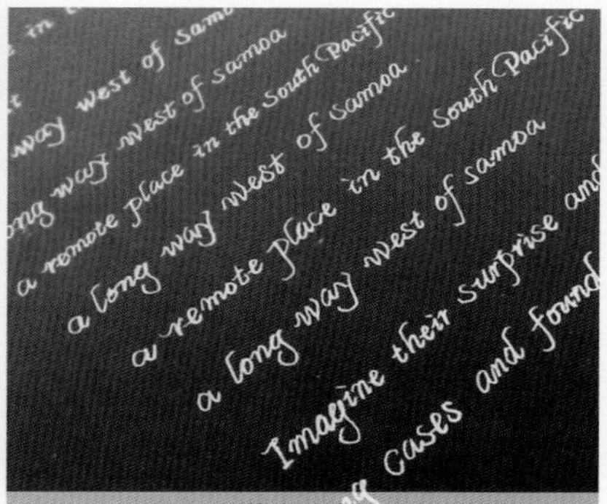

英文字符 　　　　　　　儿时的各种练习本封面 　　英语书写示范

设计灵感

高度对比色调、变化多端乐高色块给鞋底增添了层次感，引人注目的鞋带也是镶拼色配色的组成之一。
Contrasting hue, changeable Lego style color blocks and attractive shoestring add sophistication to the sole.

棉麻

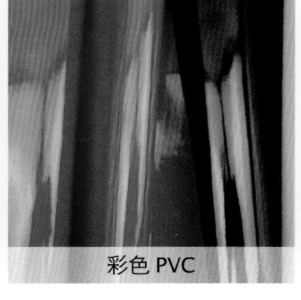

彩色 PVC

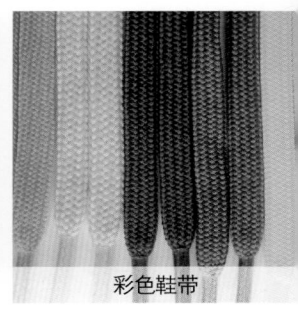

彩色鞋带

52　www.style.sh.cn

2015 SPRING/SUMMER STYLE SHANGHAI FASHION TREND / FOOTWEAR

关键要素 | 童年拾趣
KEY POINTS | CHILDHOOD REFLECTIONS

珊瑚红

石榴红

霓虹绿

湖泊蓝

迎春黄

童年学习用的英语练习册也变成了鞋子设计上有故事的元素，充满童趣而又时尚。

Elements of exercise book made for English learning can fly onto kid's shoes, making them interesting and fashionable.

泡沫橡胶底

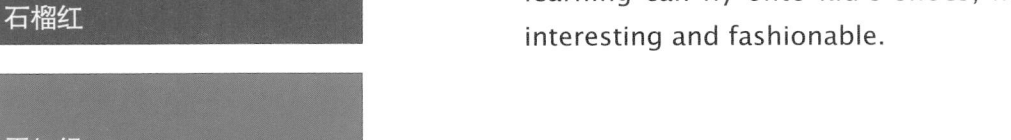

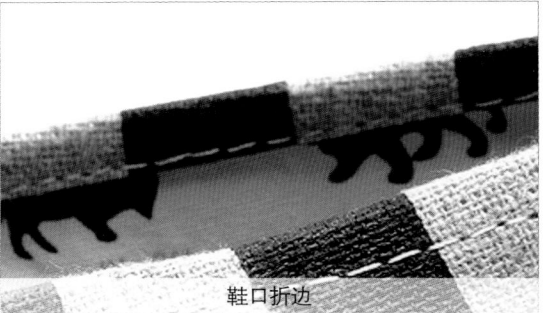
鞋口折边

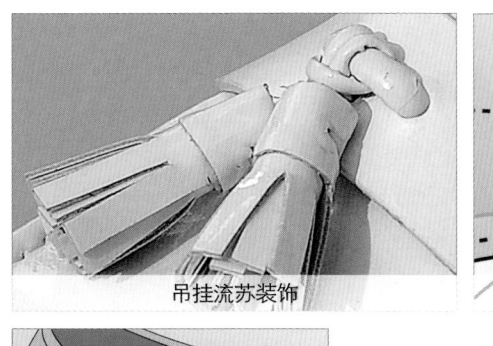
吊挂流苏装饰

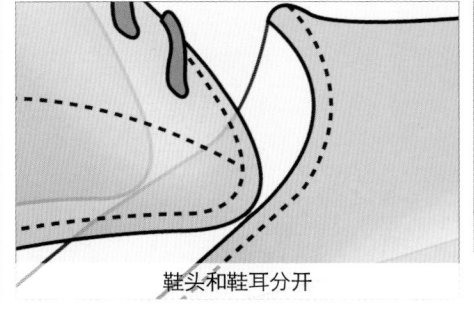
鞋头和鞋耳分开

印花和纯色布拼接

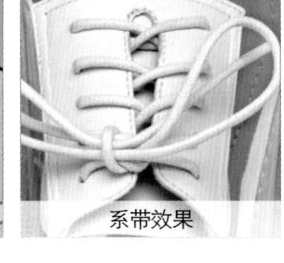
系带效果

僧侣鞋鞋头

缘自僧侣鞋的鞋型，童趣图案和原色彩色布料因对比强烈而醒目。

Prints, childish graphics and colorful cloth polish the classic monk shoes.

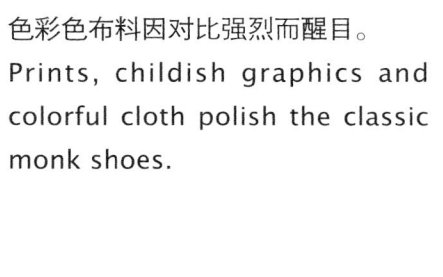
齿状鞋底线条

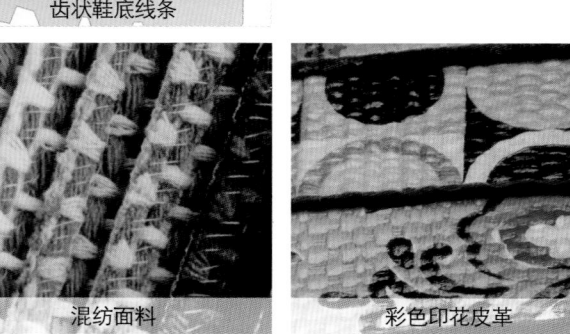
混纺面料

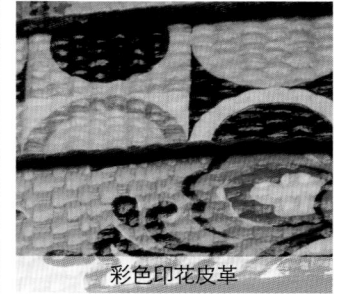
彩色印花皮革

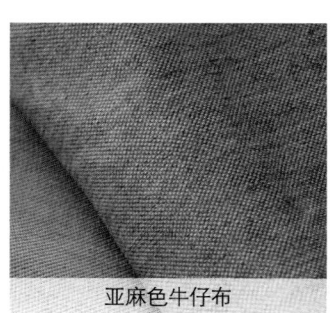
亚麻色牛仔布

透明PVC材料

绒面皮革

53

鞋履 2015 春夏海派时尚流行趋势

多彩享梦
ENJOY COLORFUL DREAMS

Normann Copenhagen & Vibeke Fonnesberg
Schmidt's 灯具设计

地毯飞行棋

地毯印花图案

Miller Goodman's 玩具

世博会某场馆外装饰墙

Britto 橱窗贴纸

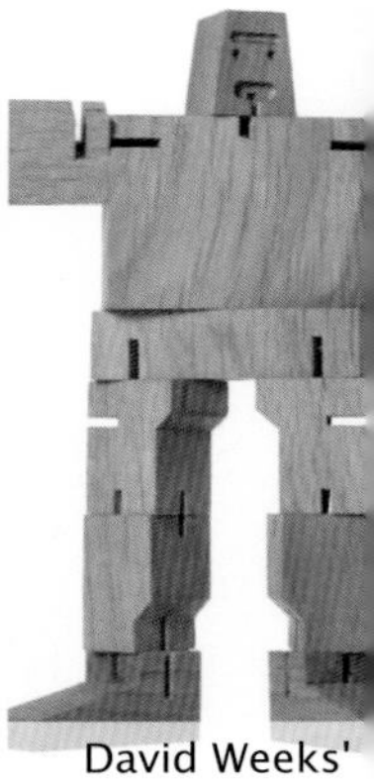

David Weeks'

设计灵感

将童年游戏元素运用在鞋靴设计中，比如用飞行棋棋盘上的各种元素打散重组运用在鞋帮面上，仿佛让我们回到了童年，而这些设计也显得更有故事。

In the footwear design, we can use elements from those games such as the flight chess and apply them on the upper. The design can easily bring us back to our childhood.

牛仔布

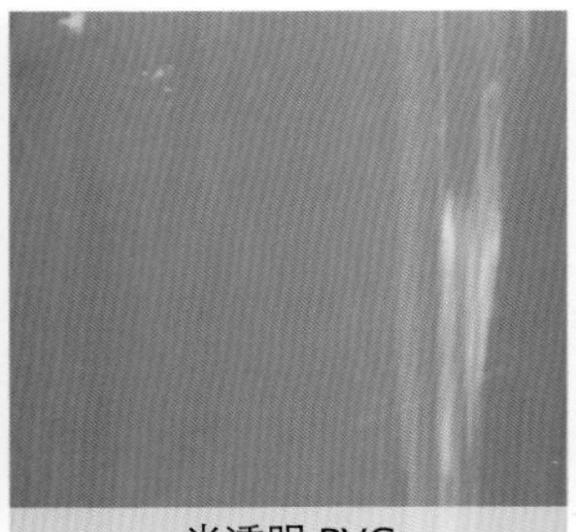

半透明 PVC

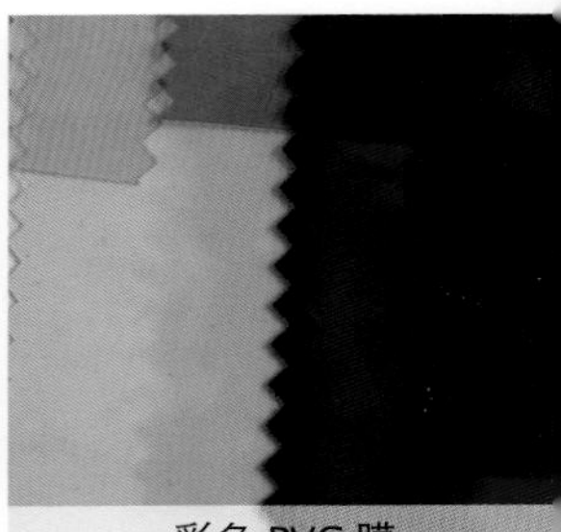

彩色 PVC 膜

54　www.style.sh.cn

2015 SPRING/SUMMER STYLE SHANGHAI FASHION TREND / FOOTWEAR

关键要素
KEY POINTS

童年拾趣
CHILDHOOD REFLECTIONS

石榴红

霓虹绿

迎春黄

琵琶橙

泳池蓝

用色块的变化、高度对比色调、变化多端的纹理给色块增添了层次感。
Colorful panels, contrasting colors and changeable textures add sophistication to our design.

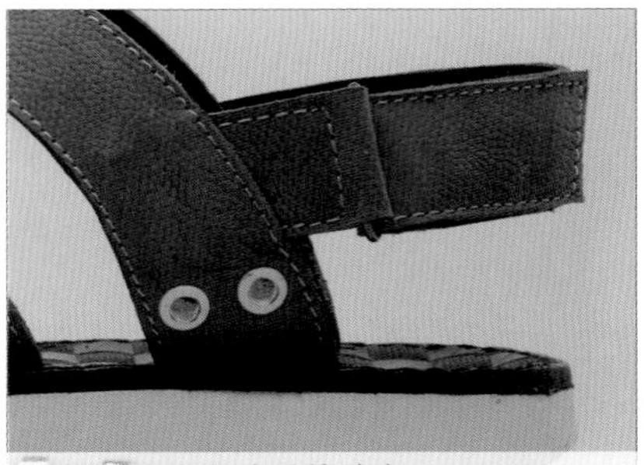

泡沫橡胶底

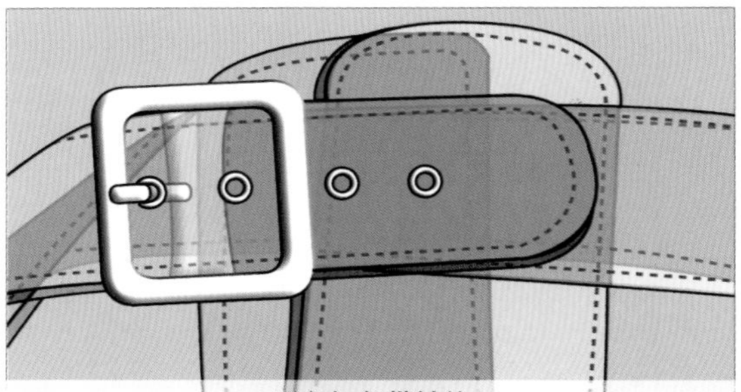

日字扣条带结构

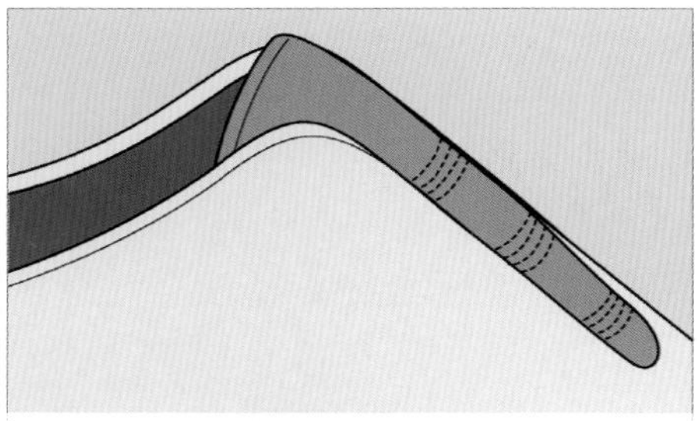

橡筋鞋舌结构

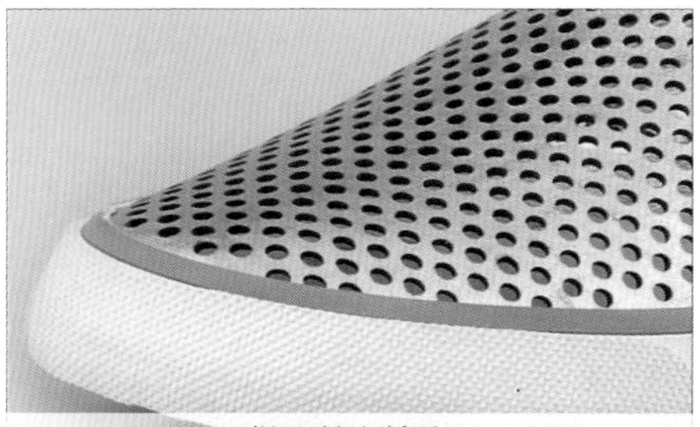

鞋面镂空结构

飞行棋变形图案

乐高积木累加效果

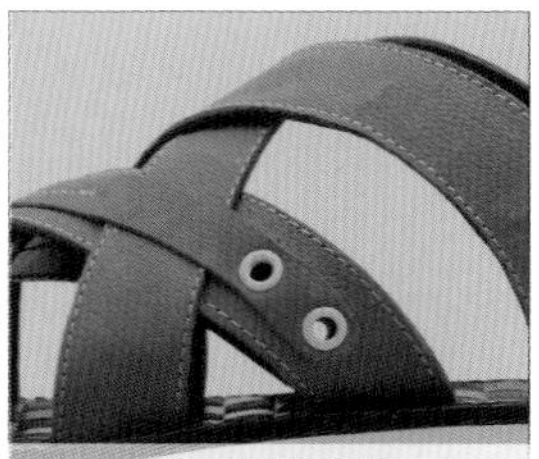

条带穿插关系

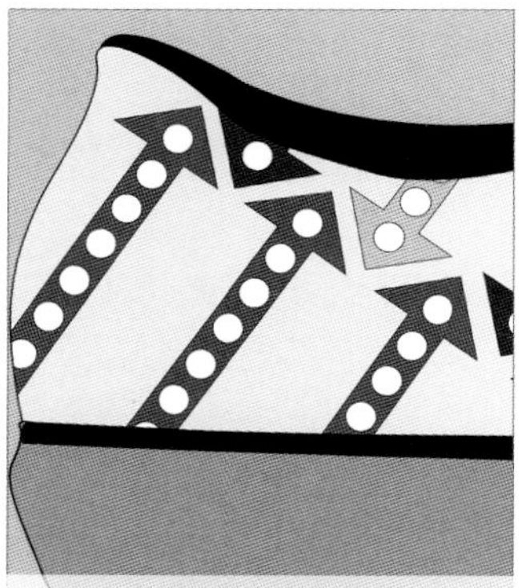

鞋面采用透明银光色 PVC 材质，还原纯真的童年情怀。
The heel looking like stacks of Lego bricks together with transparent silver PVC vamp reminds us childhood.

飞行棋棋盘变形图案

镭射高反光面料

银光色透明 PVC

魔术贴

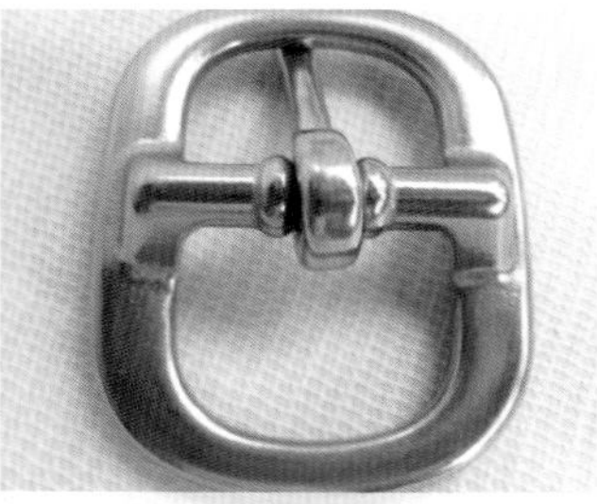

日字扣

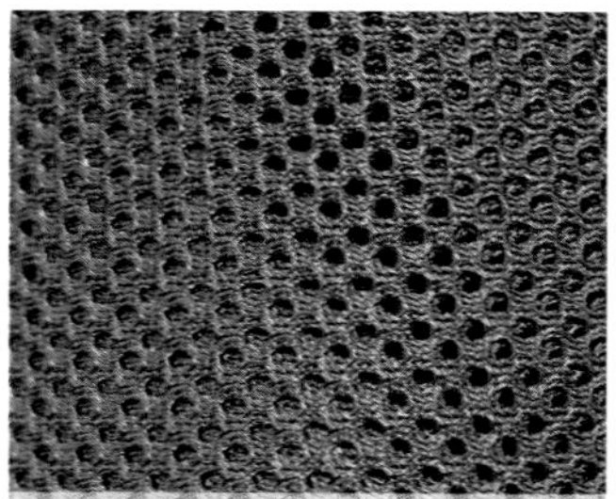

网状尼龙

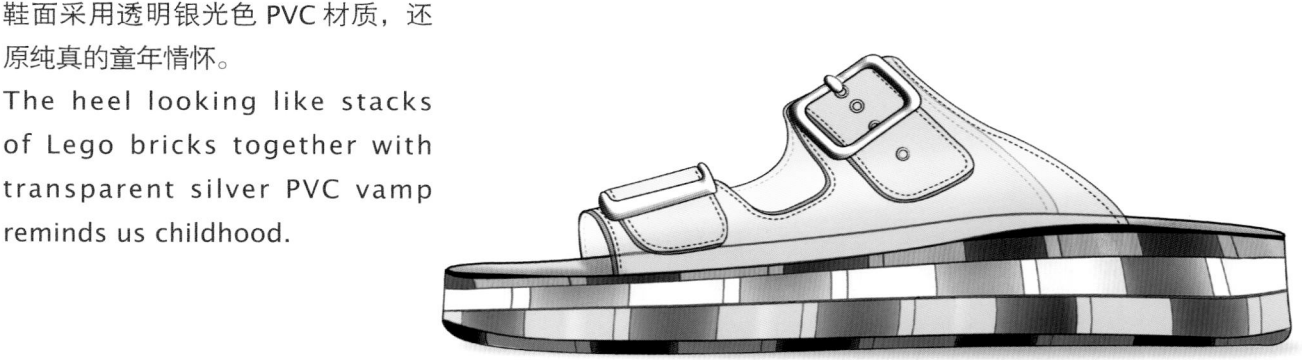
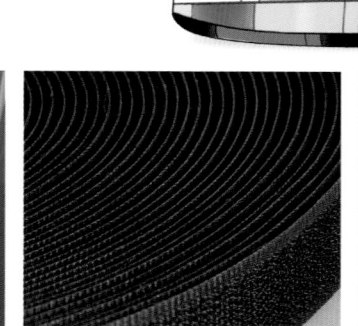
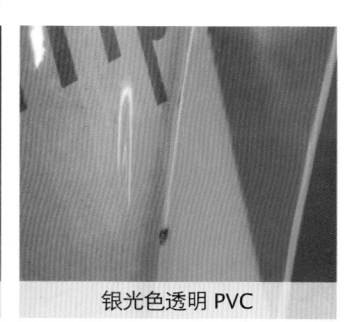
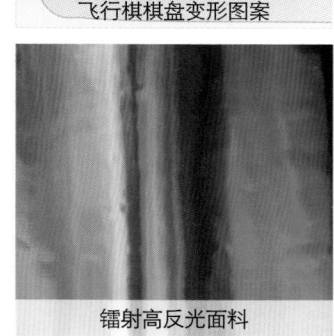

鞋履 2015 春夏海派时尚流行趋势

多彩享梦
ENJOY COLORFUL DREAMS

视觉跨界 　　　　材质的跨界 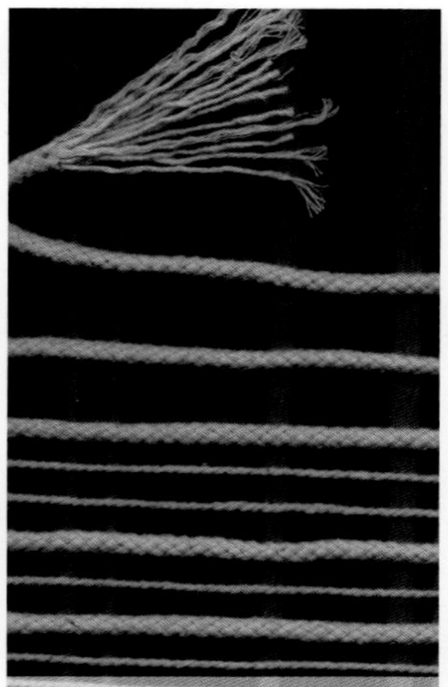　　　空间上的跨界

面料设计中透叠的层次变化　　　色彩构成 　　　雕塑上的空间变化 　　　户外立体雕塑

设计灵感

在男鞋设计中让两种形式的鞋跟同时出现在一双鞋上，造成视觉上假跟的错觉。而彩色的鞋面组合又让鞋子变得更加多彩、靓丽。

Two styles of heels on one create visual illusion and colorful vamp makes it more special.

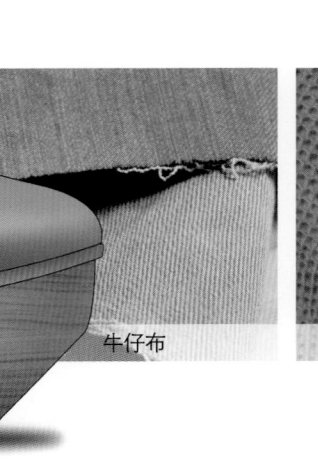

牛仔布

鱼皮

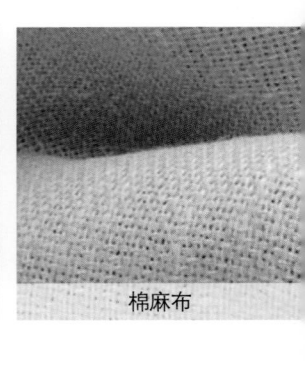

棉麻布

56　www.style.sh.cn

2015 SPRING/SUMMER STYLE SHANGHAI FASHION TREND / FOOTWEAR

关键要素 KEY POINTS | 跨界实验 TRANSBOUNDARY EXPERIMENTS

湖泊蓝

霓虹绿

松石绿

幻彩红

珊瑚红

让传统民俗的红色和绿色以比较恰当的比例组合在一起，达到一种跨界的平衡感、时尚感。
Combination of China red and bright green with appropriate proportion can be aesthetically beautiful and stylish.

绑带结构

几何视错觉图案装饰

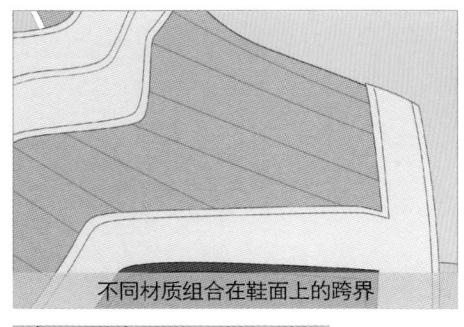

不同材质组合在鞋面上的跨界

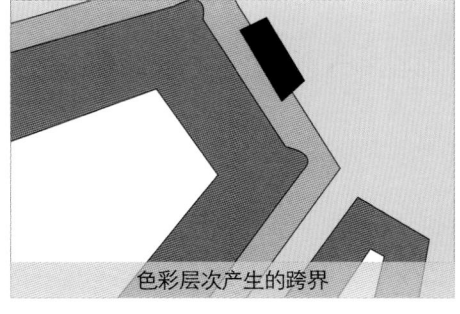

色彩层次产生的跨界

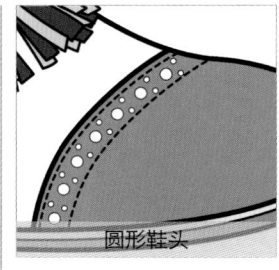

圆形鞋头

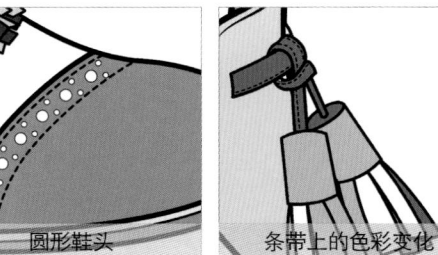

条带上的色彩变化

鞋跟线条的视觉导向

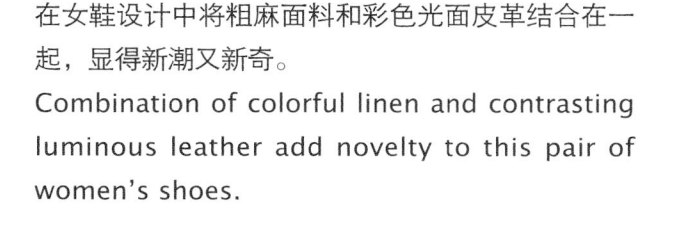

视觉假跟

在女鞋设计中将粗麻面料和彩色光面皮革结合在一起，显得新潮又新奇。
Combination of colorful linen and contrasting luminous leather add novelty to this pair of women's shoes.

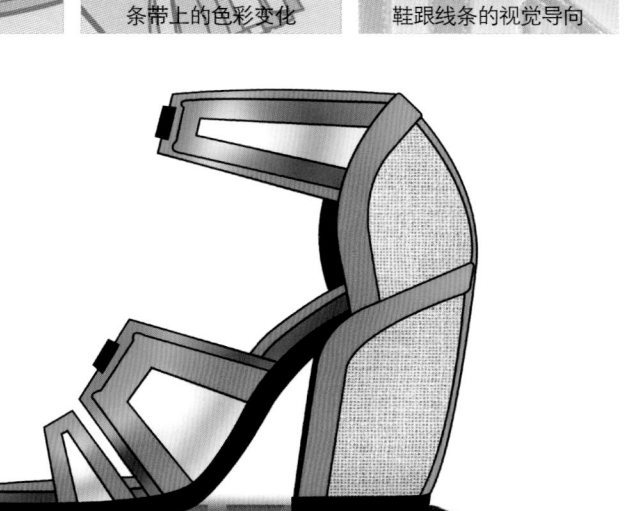

印花牛皮

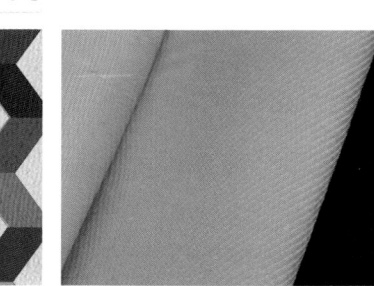

条纹布

木纹贴皮

马毛皮革

魔术贴

57

 鞋履 **2015** 春夏海派时尚流行趋势

多彩享梦
ENJOY COLORFUL DREAMS

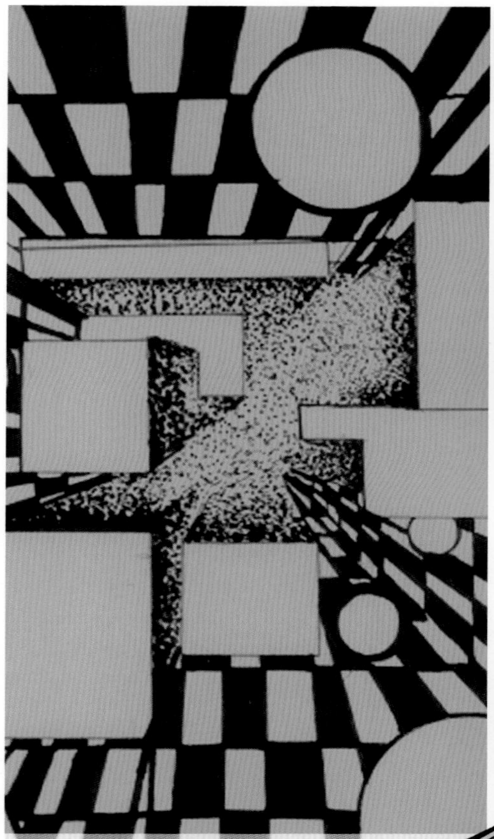

绘画艺术中的视错觉

线条粗细形成的视错觉

块面堆砌的空间视觉

空间的视错觉

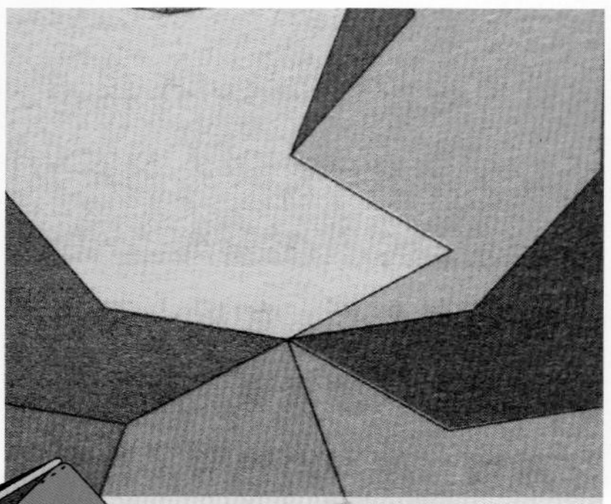

色彩拼接形成的视错觉

色彩变化形成的视错觉

方向变化的视错

设计灵感

跨界在更多的时候代表一种新锐的生活态度和审美方式，跨界音乐、跨界艺术、跨界设计、跨界营销也随之变得流行，这样更容易带来新的刺激。

Transboundary represents the new attitude to life. Cross cooperation between different industries can stimulate more new designs.

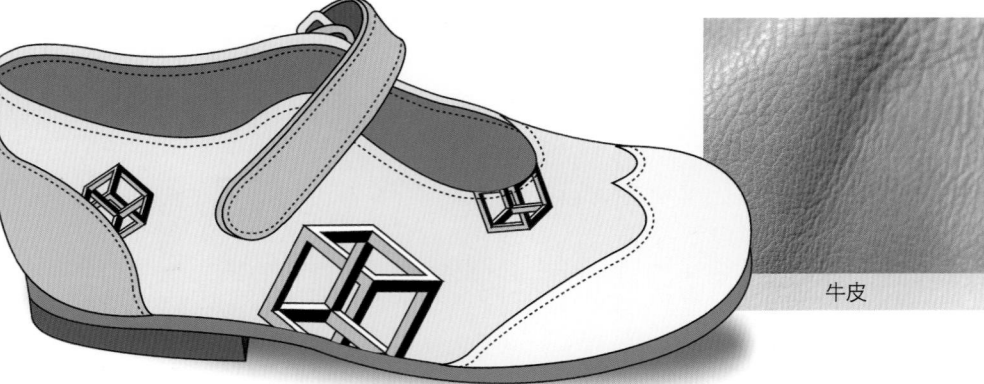

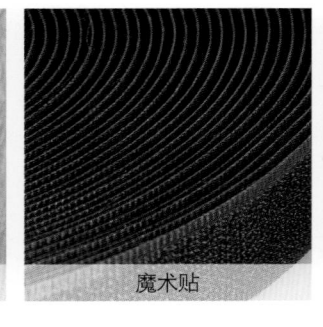

牛皮

魔术贴

各式透气网布

58　www.style.sh.cn

2015 SPRING/SUMMER STYLE SHANGHAI FASHION TREND / FOOTWEAR

关键要素 | 跨界实验
KEY POINTS | TRANSBOUNDARY EXPERIMENTS

乌梅黑

珊瑚红

迎春黄

浅草绿

泳池蓝

通过皮革的色彩变化打造立体的
错视觉效果，丰富鞋面的层次感。
Visual illusion created by
changes in color. Leather
upper adds verve.

易穿着的魔术贴结构

装饰性拉链

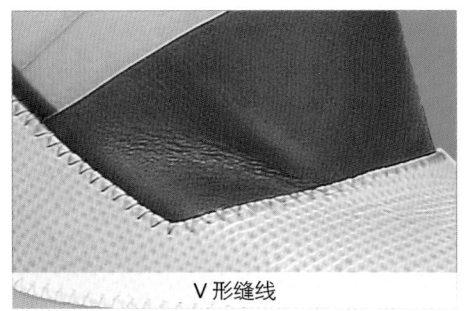

V形缝线

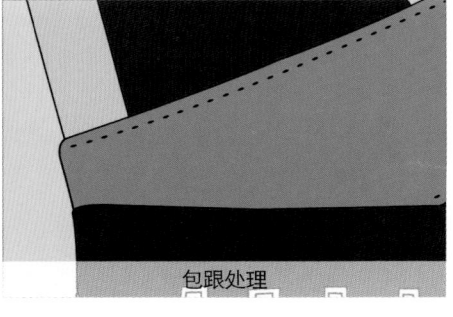

包跟处理

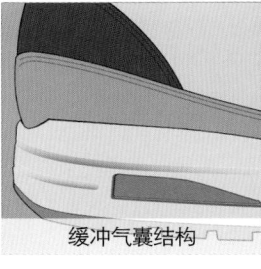

缓冲气囊结构

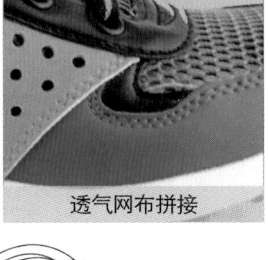

透气网布拼接

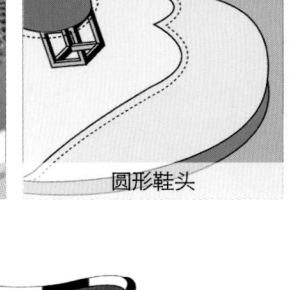

圆形鞋头

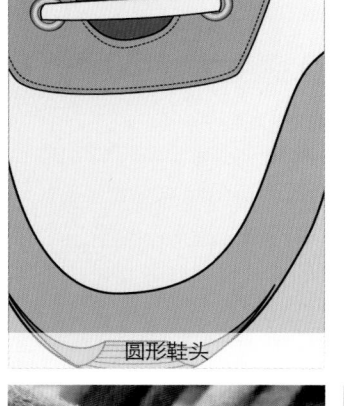

圆形鞋头

玩转几何图形，通过扭转、弯曲、
方向变化等设计方法处理几何
图形，并作为本季的印花和花
纹。
Variation of geometric
figures as add novelty to
new season collection.

生皮鞋带

尼龙鞋带

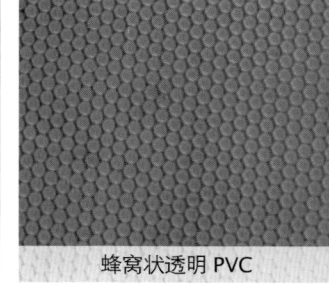

蜂窝状透明 PVC

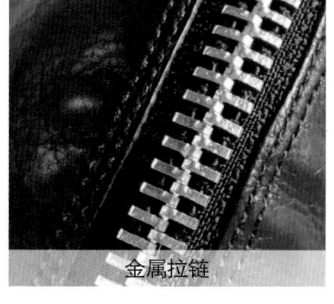

金属拉链

毛毡

59

 鞋履 2015 春夏海派时尚流行趋势

多彩享梦
ENJOY COLORFUL DREAMS

车水马龙的都市

商场外墙夜景灯光

游泳池水波纹

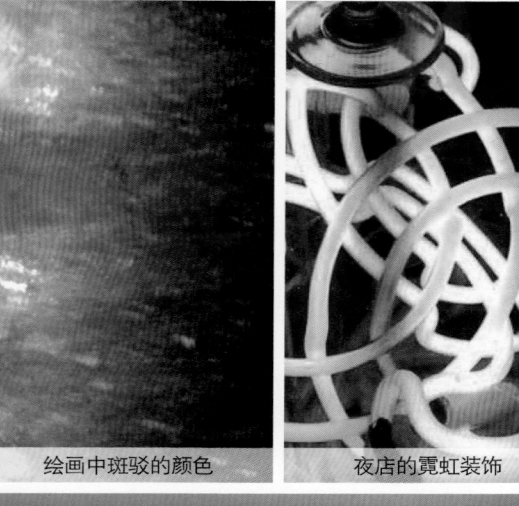

绘画中斑驳的颜色   夜店的霓虹装饰

清澈透明的海水

彩色的玻璃贝

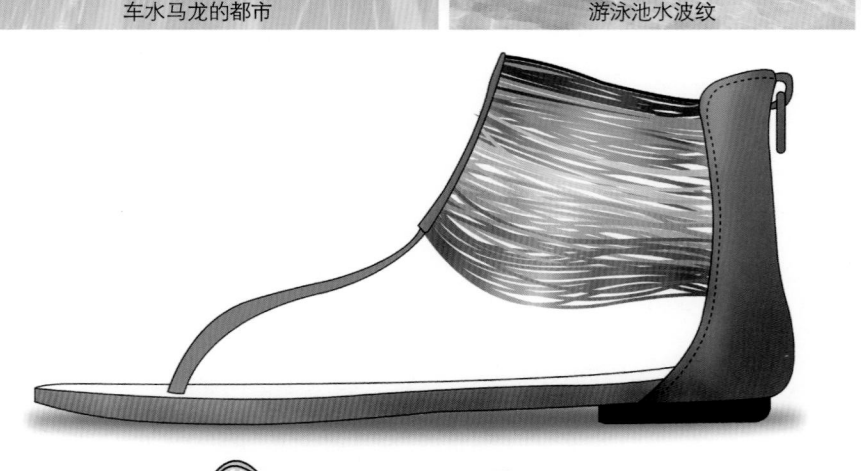

设计灵感

丰富亮丽的都市霓虹给人随性之感的色彩，走在潮流的最前端，多彩的颜色给生活带来了新的刺激。刺激过后又回到自由清新的平静状态，反映了现代生活一静一动的状态。

Face-pace city life under neon lights needs new excitement. Everything changes with times, including trends in fashion. Colorful stripes and luminescent logo brighten the design.

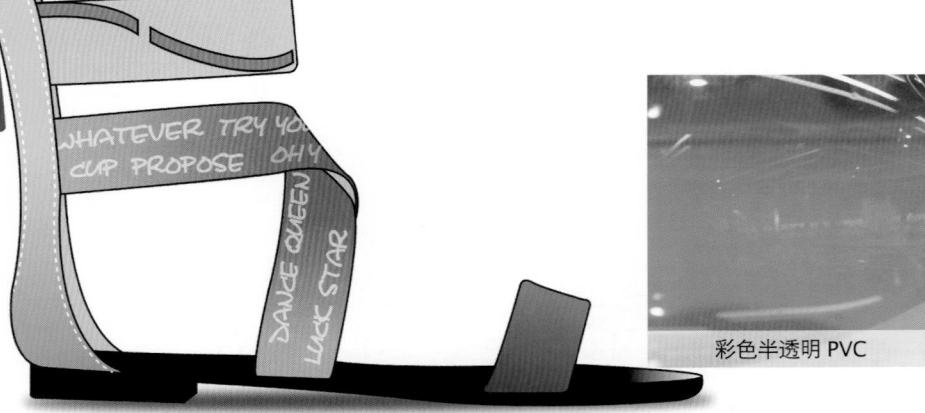

彩色半透明 PVC

镭射效果贴膜

彩色尼龙鞋带

60   www.style.sh.cn

2015 SPRING/SUMMER STYLE SHANGHAI FASHION TREND / **FOOTWEAR**

关 键 要 素 | 快乐崇拜
KEY POINTS | JOLLY IDOLATRY

木炭黑

深桃红

幻彩红

玛瑙蓝

迎春黄

鞋底和鞋面均采用印有水波纹图案的面料或者透明PVC。
Transparent PVC waviness printed sole and upper lighten the shoes.

光滑的片型中底

交织的绑带

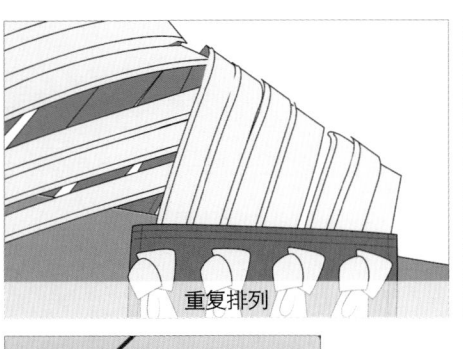
重复排列

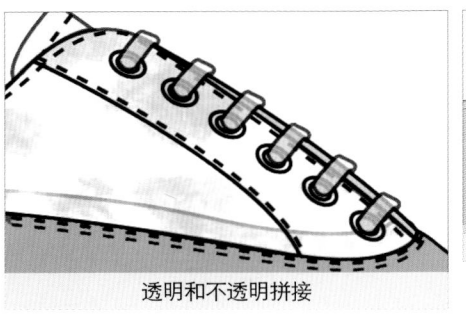
透明和不透明拼接

吊钟流苏

外发光的文字标语

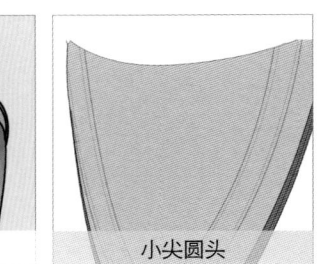
小尖圆头

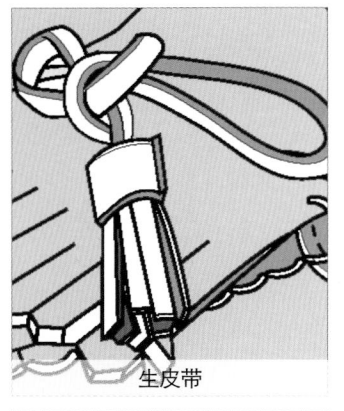
生皮带

在鞋帮面上用线性镭射高反光面料模拟水波纹的感觉。
Using lines as abstracted waves to decorate upper adds vitality to this design.

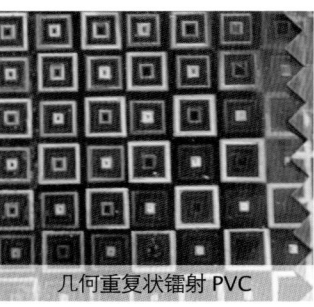
几何重复状镭射 PVC

彩色拉链头

带亮粉 PVC

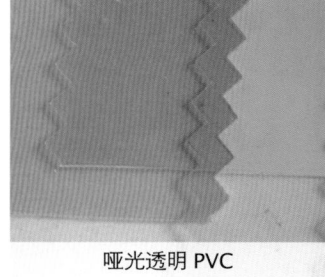
哑光透明 PVC

彩色条带

鞋履 2015 春夏海派时尚流行趋势

多彩享梦
ENJOY COLORFUL DREAMS

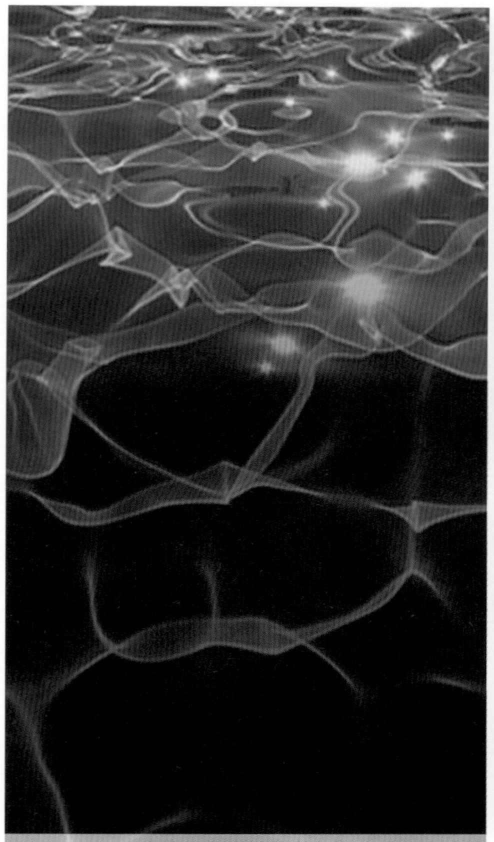

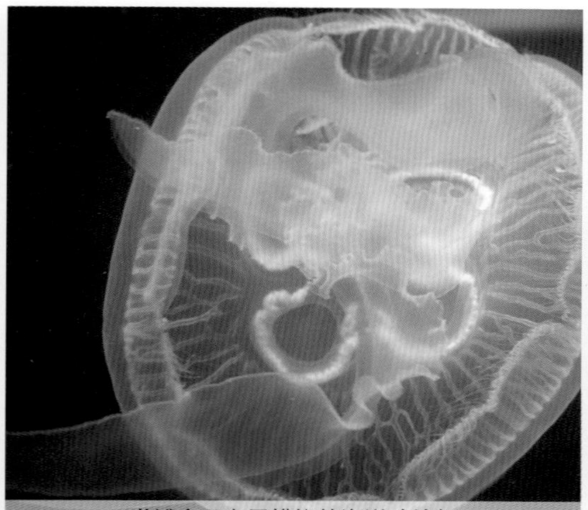

世博会里电子模拟的海洋动植物

游泳池的边缘

商场外的霓虹灯光

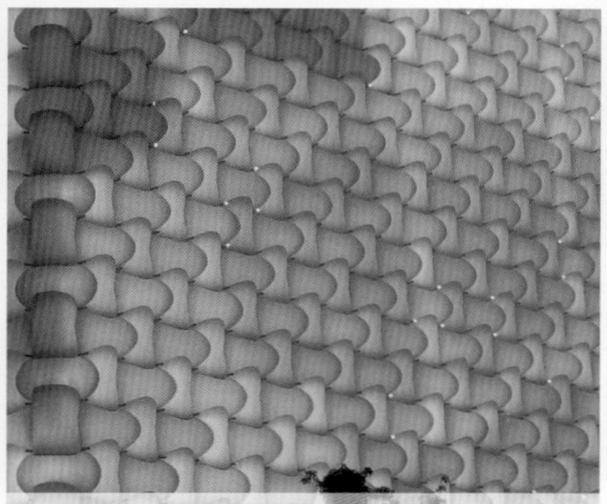

水波纹自然的色彩变化

上海世博会场馆夜景

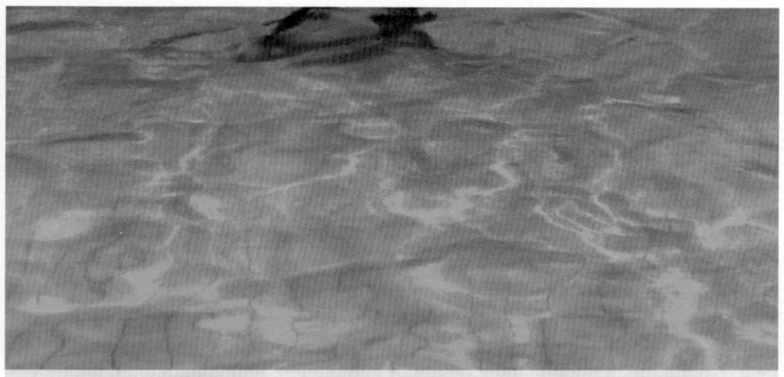

游泳池里清澈的波纹

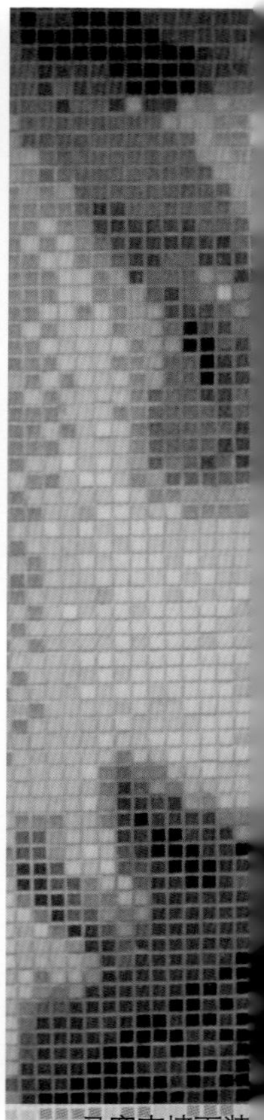

马赛克墙面装

设计灵感

"让我们荡起双桨，小船儿推开波浪 …… 阳光洒在海面上"，将这样荡漾着的水纹以摇动的姿态出现在鞋履设计中，让意境具象化，仿佛带我们到了水边，让我们感受到纯粹平静的快乐。

"Let's sway twin oars, Our little boat's plowing across the ripples", "And so striking in sunlight are red scarves", these lyrics depict a beautiful scene. Inspired by this, we highlight the design with waviness.

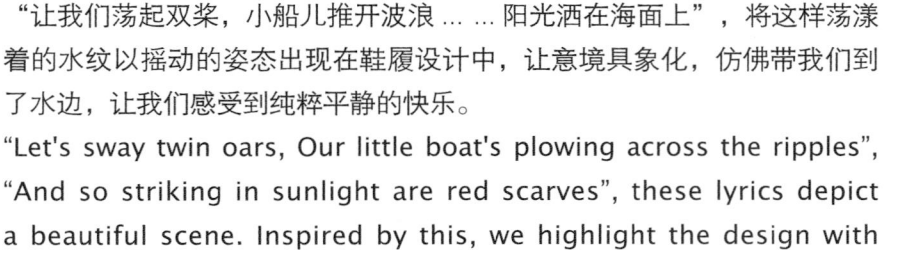

印有水波纹的牛仔布

透气网布

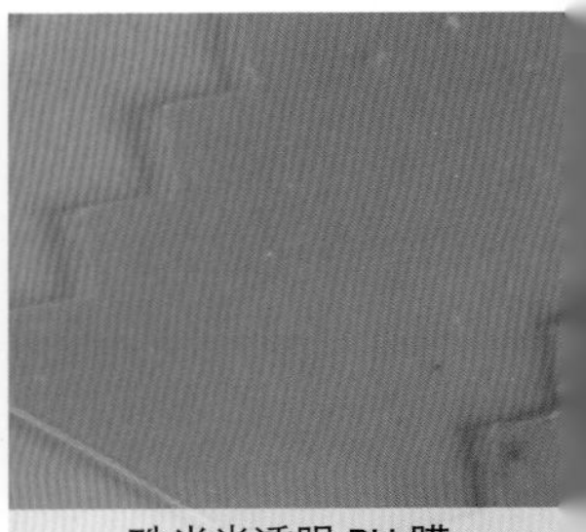

珠光半透明 PU 膜

62   www.style.sh.cn

2015 SPRING/SUMMER STYLE SHANGHAI FASHION TREND / **FOOTWEAR**

关 键 要 素 | 快乐崇拜
KEY POINTS | JOLLY IDOLATRY

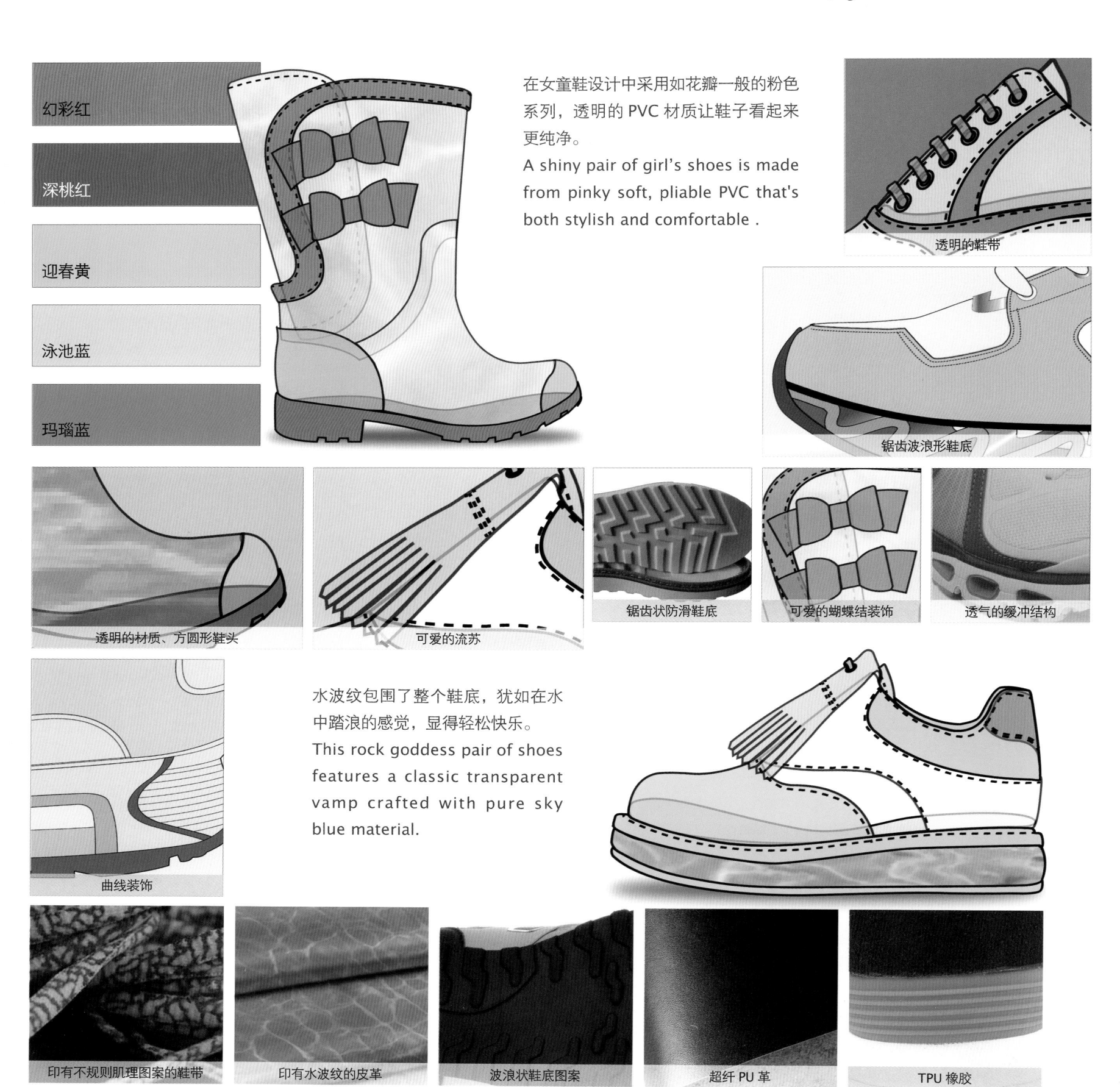

幻彩红

深桃红

迎春黄

泳池蓝

玛瑙蓝

在女童鞋设计中采用如花瓣一般的粉色系列，透明的 PVC 材质让鞋子看起来更纯净。
A shiny pair of girl's shoes is made from pinky soft, pliable PVC that's both stylish and comfortable .

透明的鞋带

锯齿波浪形鞋底

透明的材质、方圆形鞋头

可爱的流苏

锯齿状防滑鞋底

可爱的蝴蝶结装饰

透气的缓冲结构

水波纹包围了整个鞋底，犹如在水中踏浪的感觉，显得轻松快乐。
This rock goddess pair of shoes features a classic transparent vamp crafted with pure sky blue material.

曲线装饰

印有不规则肌理图案的鞋带

印有水波纹的皮革

波浪状鞋底图案

超纤 PU 革

TPU 橡胶

63

S 鞋履 2015 春夏海派时尚流行趋势

多彩享梦
ENJOY COLORFUL DREAMS

英文字母图形　　街头艺术　　超人"连环画片段

《美国队长》连环画片段　　涂鸦英文字　　M50 创意园区涂鸦墙　　涂鸦主题的平面

设计灵感

都市街头的男鞋受到现代时尚、街头艺术等影响变得更加张扬外放，通过印花和间隔印染的方式，结合明亮、刺激的色彩，彰显夏季时光的活力，将一种全新的正能量传递给消费者。

Men's shoes nowadays have been more and more influenced by fashion industry and street culture. Bold bright and eye-catching colored prints make add vitality to them.

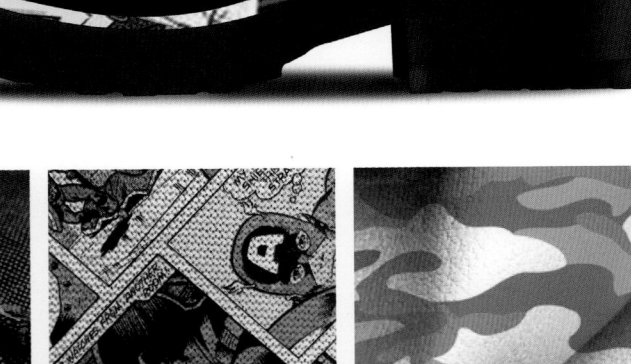

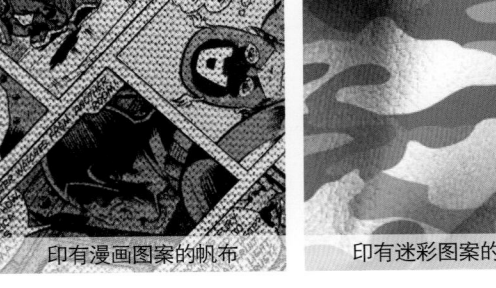

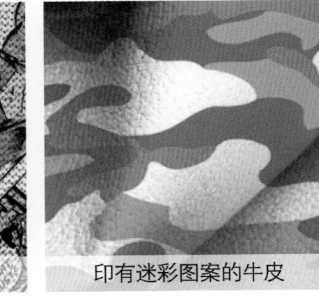

纯色牛仔布　　印有漫画图案的帆布　　印有迷彩图案的牛皮

64　www.style.sh.cn

2015 SPRING/SUMMER STYLE SHANGHAI FASHION TREND / **FOOTWEAR**

关 键 要 素 │ 都市街头
KEY POINTS │ CITY STREETS

乌梅黑

迎春黄

泳池蓝

玛瑙蓝

石榴红

黑色的鞋帮面皮料组合涂鸦型鞋内里，体现了现代都市人在生活压力下渴望释放解压的心态。

A black leather wrapped wedge with graffiti printed insole furthers the hip, bohemian style of a colorful sneaker.

粗犷生皮鞋带

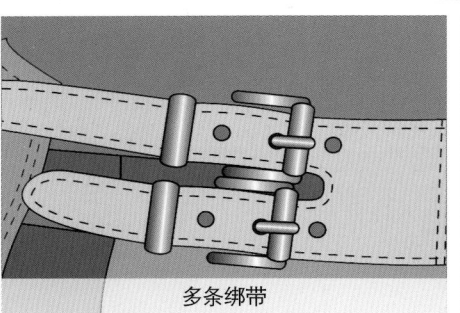

多条绑带

厚重的坡跟

纹理或波纹状鞋底

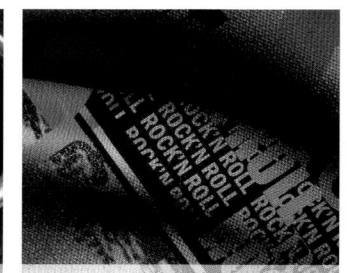

厚防水台

锯齿形鞋底造型

宽圆形鞋头

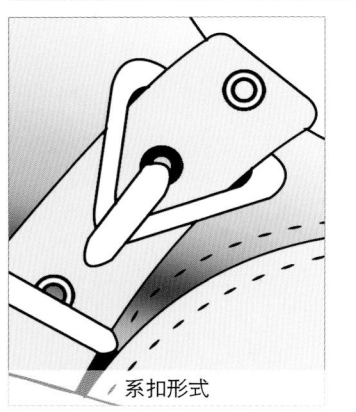

系扣形式

采用鲜明的亮色涂鸦图案和纯色皮面或透明鞋面结合尽显都市街头风尚。

An exotic bright graffiti print marks a sleek classic clad in leather or transparent vamp.

橡胶鞋底

古铜色钥匙圈、弹簧圈

印花牛津布

彩色亚克力

印花帆布

65

鞋履 2015 春夏海派时尚流行趋势

多彩享梦
ENJOY COLORFUL DREAMS

大师油画中的都市街头

图形化的电梯门装饰

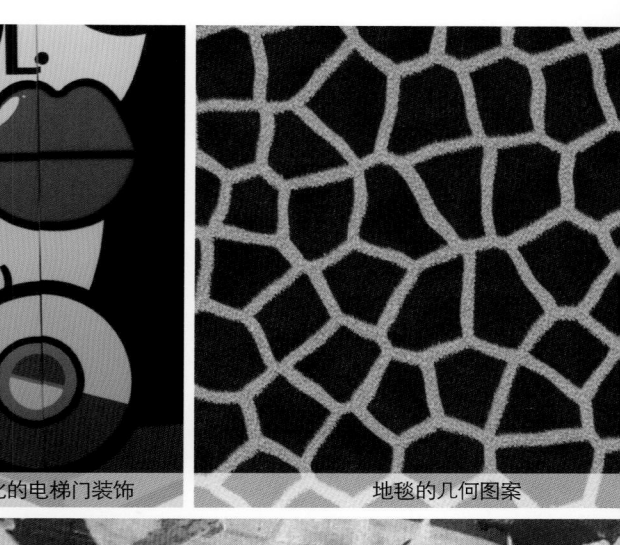
地毯的几何图案

涂鸦艺术中跳跃的点线面构成

油画中块面的色彩分割

喷溅涂鸦效果

设计灵感

都市街头的童鞋采用街头涂鸦艺术中明亮的色彩对比、明亮的色彩镶拼或荧光色细节让儿童鞋变得更加可爱，在鞋面局部混搭动物皮纹点缀，充当中性新元素。

Bright color graffiti, exotic leopard-spot or lunar prints on kid shoes or sneakers make them more adorable.

反绒牛皮

涤纶布

毛毡

www.style.sh.cn

2015 SPRING/SUMMER STYLE SHANGHAI FASHION TREND / FOOTWEAR

关键要素 | 都市街头
KEY POINTS | CITY STREETS

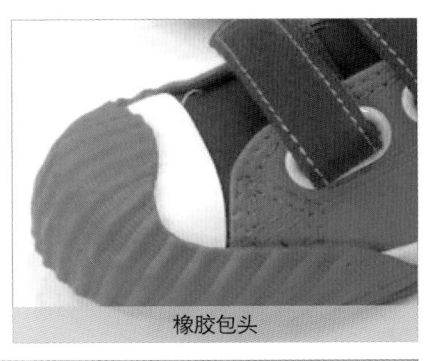
橡胶包头

乌梅黑

玛瑙蓝

珊瑚红

迎春黄

石榴红

坡跟将运动鞋变得端庄、时尚，其高帮结构 、彩色的图案 和魔术贴结构显得复古。
Stand out from the crowd in latest wedge sneaker, finished with swirly prints. Use the Veltro straps to perfect the fit.

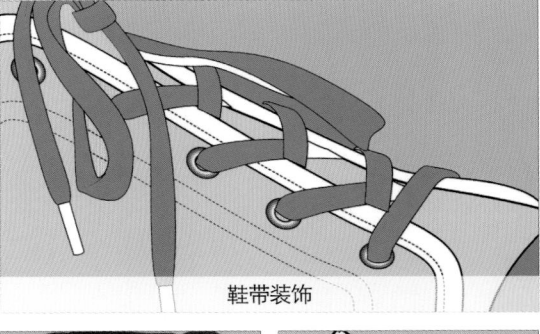
鞋带装饰

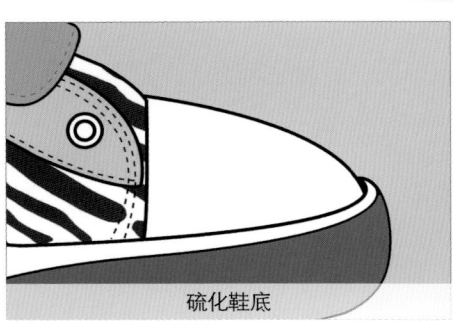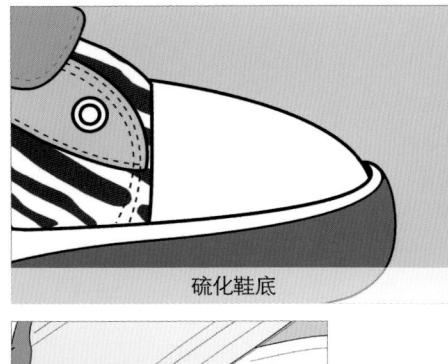
硫化鞋底

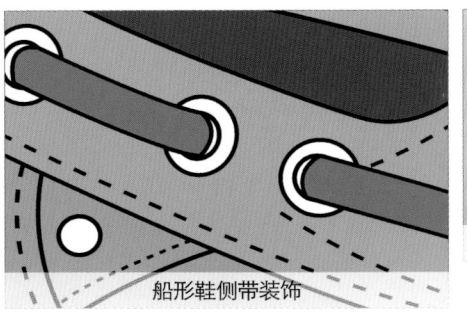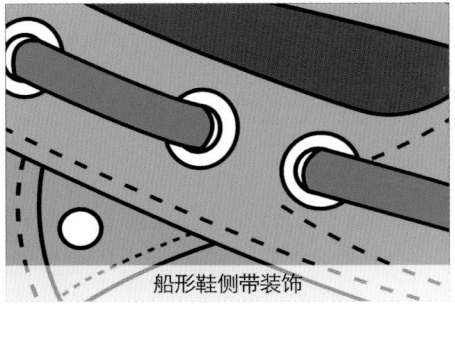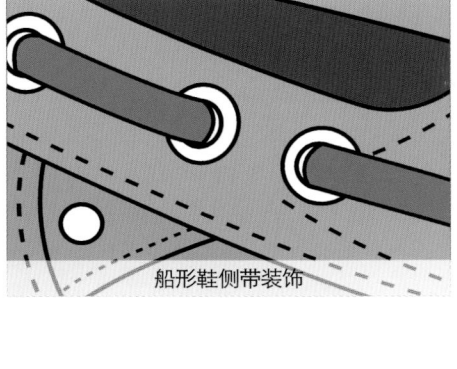
船形鞋侧带装饰

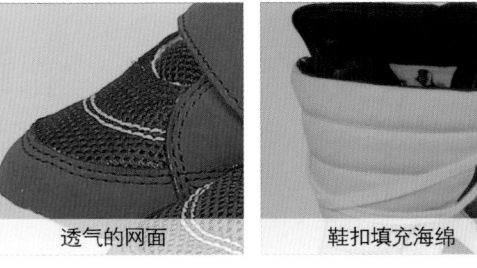
透气的网面

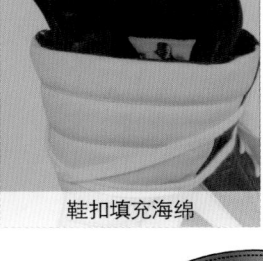
鞋扣填充海绵

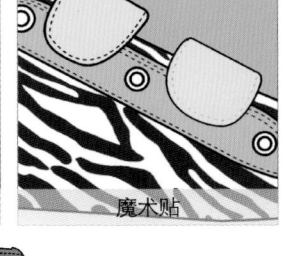
魔术贴

运动坡跟鞋以不同的内增高度、不同的线条穿插效果、繁复图案、鲜艳多彩的亮色成为新的亮点。
Color-pop accents further the hip, trend-right style of hidden-wedge sneakers created in different heights and colorful prints.

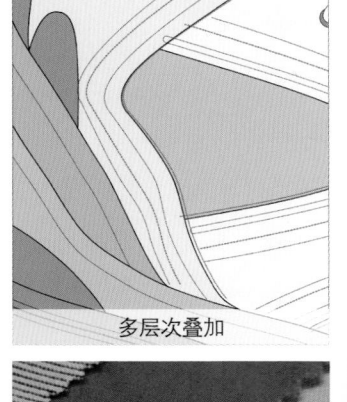
多层次叠加

彩色梭织布

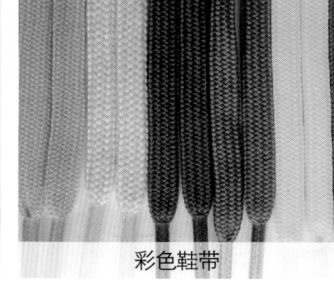
彩色鞋带

印花皮革

魔术贴

印花/压印皮革

67

68　www.style.sh.cn

箱包 BAG&SUITCASE

海上星梦
STAR DREAMS OVER THE SEA

优活之梦
LIVING THE DREAM

时空筑梦
REALISING DREAMS THROUGH SPACE

多彩享梦
ENJOY COLORFUL DREAMS

S 箱包 2015 春夏海派时尚流行趋势

海上星梦
STAR DREAMS OVER THE SEA

上海以其独特的魅力吸引着无数追逐梦想的年轻人。这些年轻人来这里学习、工作，并时刻为这座城市注入新的能量与活力，他们对于生活充满激情，他们享受生活，同样也为生活而努力奋斗。

Shanghai with her unique charm has attracted countless young people to study or work here, injecting energy and vitality into the city. They are passionate about life and willing to work hard for better life.

BAG&SUITCASE
2015 SPRING/SUMMER STYLE SHANGHAI FASHION TREND
海派箱包流行趋势

2015 SPRING/SUMMER STYLE SHANGHAI FASHION TREND / **BAG&SUITCASE**

关键要素 | KEY POINTS

繁华时代
FLOURISHING ERA

精细品质
EXQUISITE QUALITY

新贵气质
NOBLE TEMPERAMENT

创新传统
INNOVATING TRADITIONS

青釉灰

青砖灰

杏仁黄

皮革黄

琥珀橙

酸枝红

紫檀棕

瓷瓦黑

铁艺灰

青瓷灰

S 箱包 2015 春夏海派时尚流行趋势

海上星梦
STAR DREAMS OVER THE SEA

夜晚的玻璃橱窗 　　旗袍盘扣 　　绣花鞋鞋面

视错觉灯光　　新型细线灯管 　　复古宝石头饰　　复古建筑装饰

设计灵感

印花贴图和经典女包款式结合，运用金属配件，彰显包袋的高档大气，低调奢华。

A printed finish refines a structured tote accented with enameled logo hardware for added signature appeal.

黑色皮料　　雾面金葱布料　　传统印花布

72　www.style.sh.cn

2015 SPRING/SUMMER STYLE SHANGHAI FASHION TREND / BAG&SUITCASE

关键要素 | 繁华时代
KEY POINTS | FLOURISHING ERA

琥珀橙

杏仁黄

紫檀棕

浪花白

乌梅黑

童包上绣上亮片，配合珍珠搭扣和链条提带。

A sweet kid bag sparkles with a multi hued glittery finish, a pearl-topped closure and metallic chains.

珍珠垂坠

大面积绣布

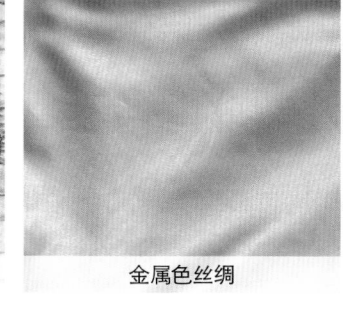

金属链条和花型装饰

金属扣

口金扣

锁型链条

金属包扣

水晶扣

女士手包，以亮色的蛇皮、蟒蛇皮或牛皮压花体现贵气繁华之感，配合金属质感的配饰，营造大气繁华的经典。

Metallic luster gleams atop handbags crafted from luminescent emossing snakeskin, pythonskin or calfskin leather while metallic hardware details those bags for a polished finish.

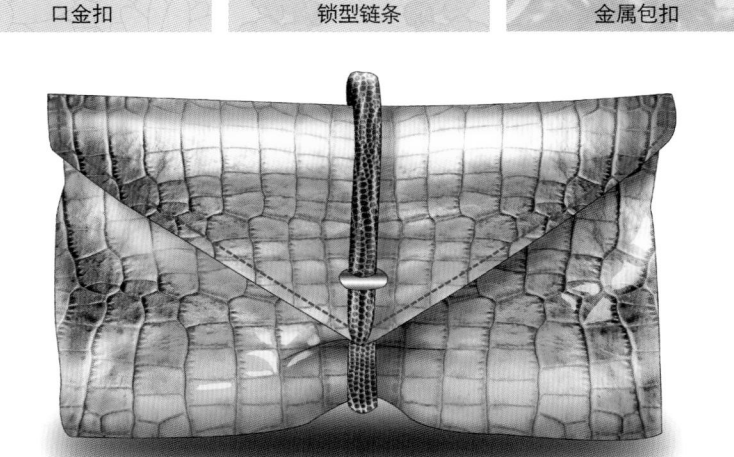

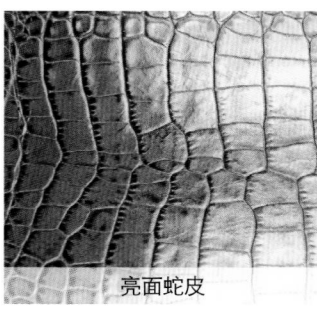

亮面蛇皮

金属色丝绸

金葱饰布

水晶饰布面料

印花面料

 上海设计之都
公共服务平台
认定专业平台

 STYLE SHANGHAI

73

S箱包 2015 春夏海派时尚流行趋势

海上星梦
STAR DREAMS OVER THE SEA

旗袍人型挂台

柜台项链展示

留声机

精致瓷盘

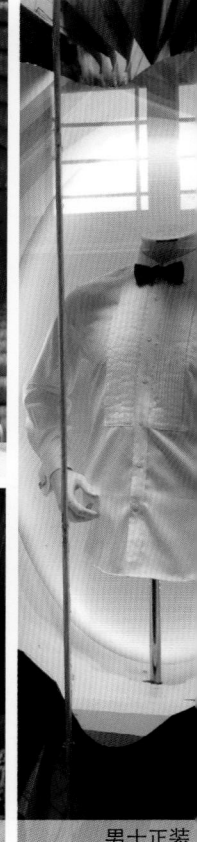
男士正装

老式黑胶碟

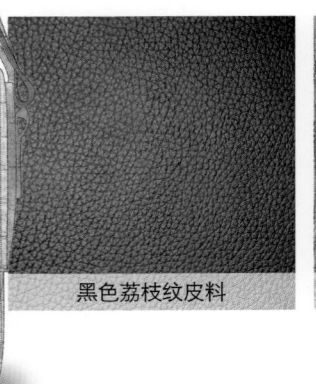
饰品展柜

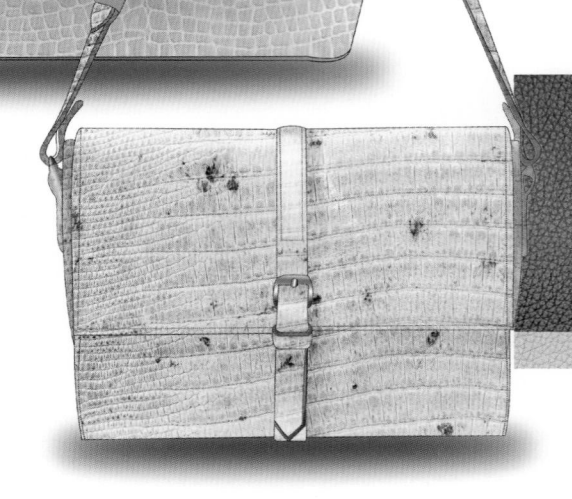

设计灵感

以皮料为主，经典简约的包款，搭配整块的色块皮料和金属配件，大气精致又不沉闷老气，结合春夏色调整体高档经典。

Supple calfskin leather defines a clean, sophisticated tote accented with glimmering gold tone hardware.

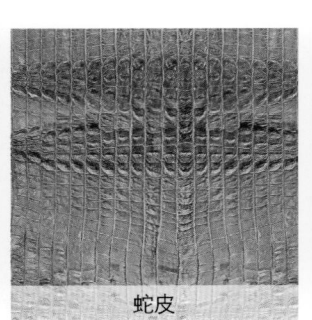

黑色荔枝纹皮料　　　蛇皮　　　红色牛皮

74　www.style.sh.cn

2015 SPRING/SUMMER STYLE SHANGHAI FASHION TREND / BAG&SUITCASE

关键要素 | 精细品质
KEY POINTS | EXQUISITE QUALITY

杏仁黄

皮革黄

豆浆白

浪花白

酸枝红

蟒蛇皮和牛皮的拼接，在原本经典的款式上增加了春夏的活力，也不失简约经典。

Creative design brings signature shine to this roomy, relaxed bag in an indulgent mix of tactile calfskin and snake-embossed leather.

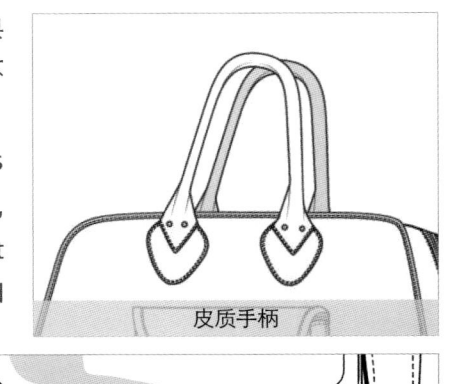

皮质手柄

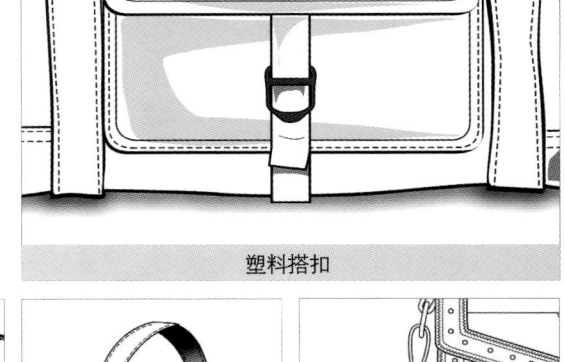

塑料搭扣

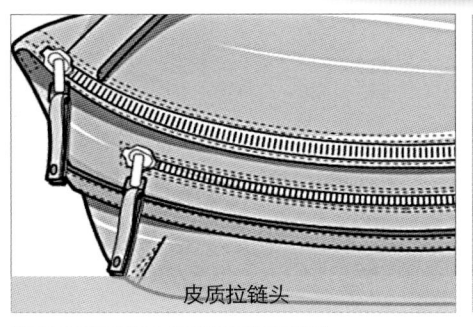

皮质拉链头

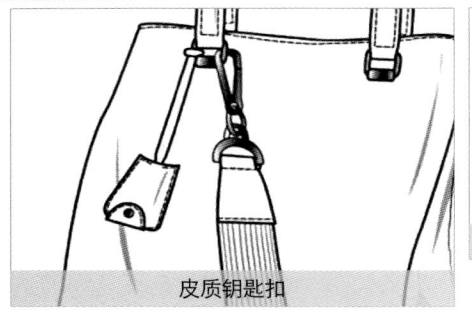

皮质钥匙扣

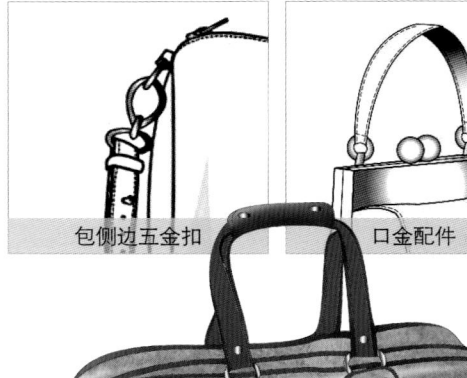

包侧边五金扣

口金配件

皮质气孔装饰

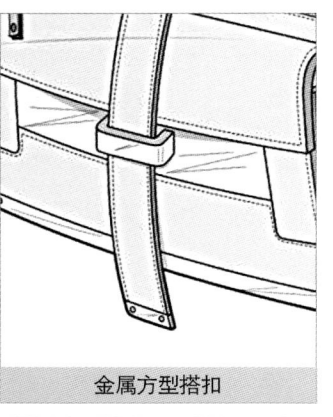

金属方型搭扣

雾面反光皮质感的男士大斜跨拎包，结合三种不同的皮料质感，大气精致。

Three different kinds of supple leather structure a stylish messenger bag featuring a sleek silhouette and urban sensibilities.

蟒蛇皮

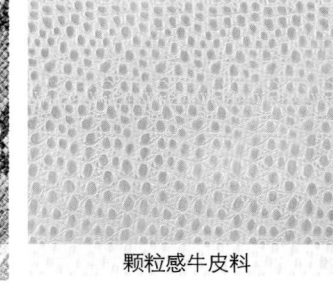

颗粒感牛皮料

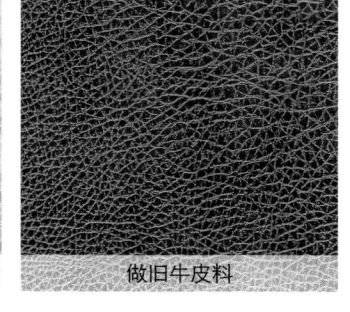

做旧牛皮料

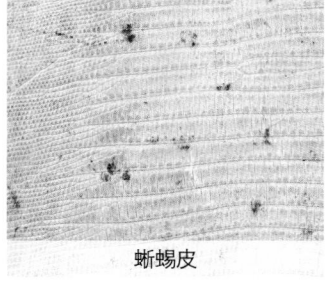

蜥蜴皮

咖啡色蟒蛇皮

75

S 箱包 2015 春夏海派时尚流行趋势

海上星梦
STAR DREAMS OVER THE SEA

橱窗首饰展示

精致小店

精致的餐桌

手工藤椅

贴钻的玻璃镜

酒瓶排列展示

百货内室装潢

设计灵感

皮质公文包和拉杆箱，质的追求，包边的细节处理，有份量的金属扣，体现男士们对品质的追求。

Buckle-detailed briefcases or carry-on perfecting functional fashion with supple leather exteriors are always essentials.

淡色蛇皮　　金色蛇皮　　反绒皮

76　www.style.sh.cn

2015 SPRING/SUMMER STYLE SHANGHAI FASHION TREND / BAG&SUITCASE

关键要素 | 精细品质
KEY POINTS | EXQUISITE QUALITY

咖啡棕

皮革黄

杏仁黄

剑麻灰

青砖灰

皮质小手提箱包，细巧精致感，滚边不同色块皮质的拼接，精细不沉闷。
Silky leather patches upgrade a capacious top-handle carry-on perfect for a full day or weekend getaway.

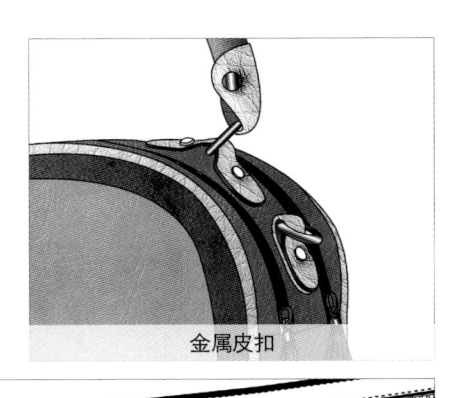
金属皮扣

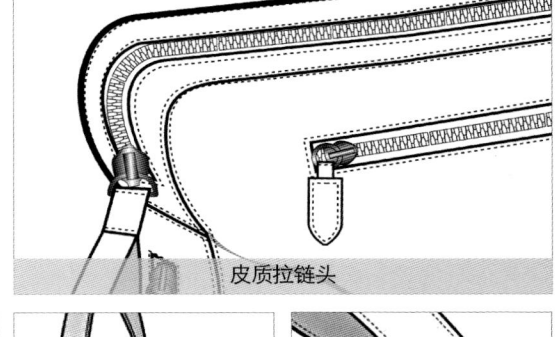
皮质拉链头

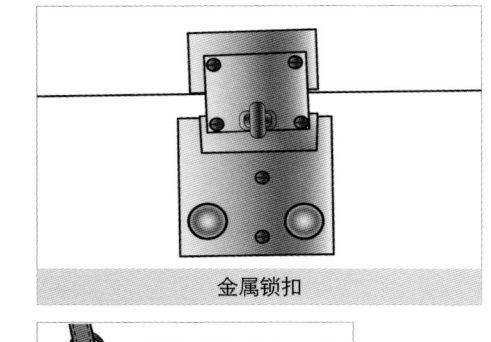
金属锁扣

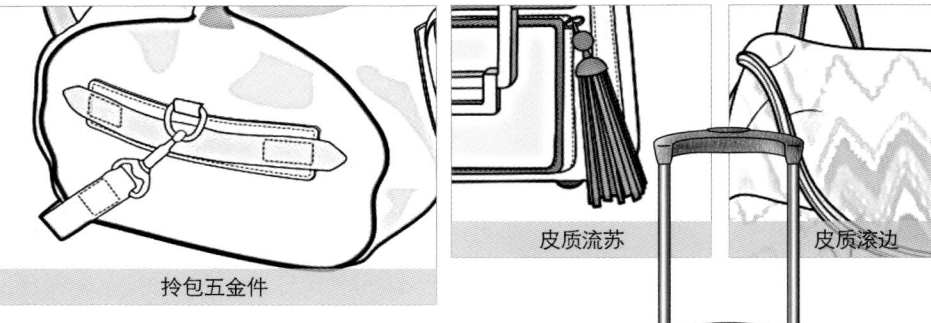
拎包五金件

皮质流苏

皮质滚边

斜挎搭扣五金

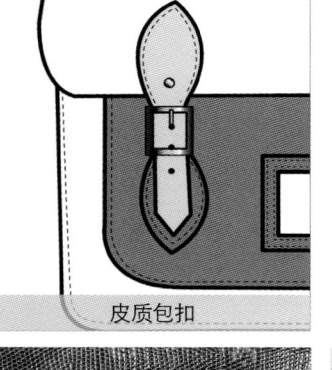
皮质包扣

皮质感给人以高端大气精致生活的感觉，皮质感的箱包耐用而且体现品味。
"Quality is not an act, it is a habit." A leather bag starts it all.

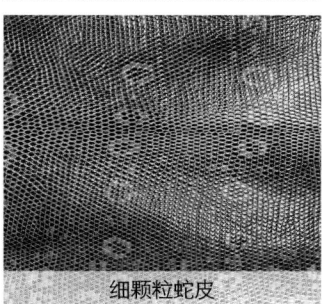
细颗粒蛇皮

毛面蛇皮

压花牛皮

猪皮

啡色牛皮

上海设计之都
公共服务平台
认定专业平台

STYLE SHANGHAI

77

箱包 2015 春夏海派时尚流行趋势

海上星梦
STAR DREAMS OVER THE SEA

丝绸质丝巾 　　下午茶餐具 　　走廊灯光

外滩夜景　　网格窗 　　欧式餐盘　　层叠建筑装饰

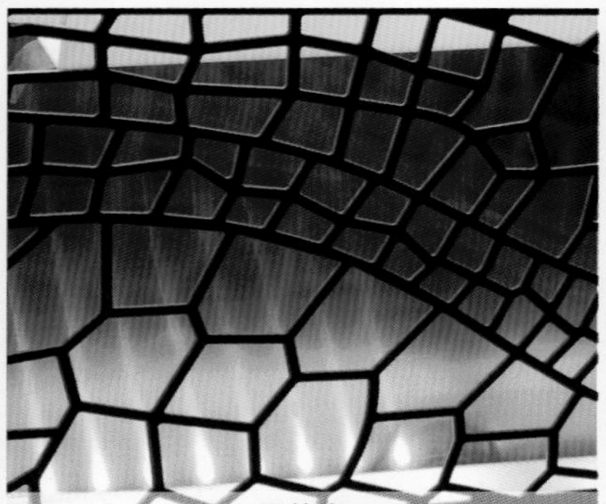

设计灵感

以欧式印花为主题的新贵女包，印花布面配合牛皮的质感和英式的气孔以及皮质流苏，营造整体新贵的感觉。

The leather tassel adds new dimension to the essential bag with intricate prints and engraving, making it a modern essential.

印花条纹布　　红牛皮　　印花布

78　www.style.sh.cn

2015 SPRING/SUMMER STYLE SHANGHAI FASHION TREND / BAG&SUITCASE

关 键 要 素 | 新贵气质
KEY POINTS | NOBLE TEMPERAMENT

咖啡棕

麦糖黄

胭脂红

浪花白

豆浆白

男式手拎包，经典欧式的款式，配上质感的牛皮，加上欧式花纹。
This classic getaway bag features elegant prints and supple leather.

气孔

皇冠图案

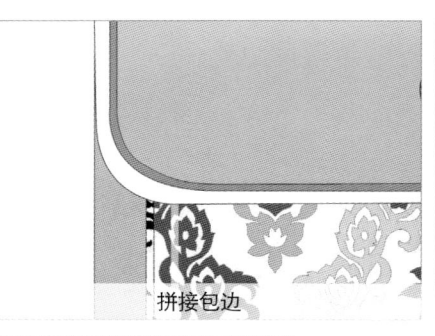

拼接包边

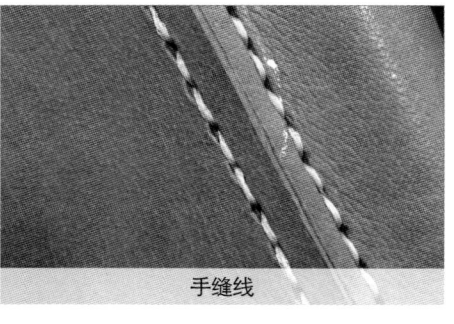

手缝线

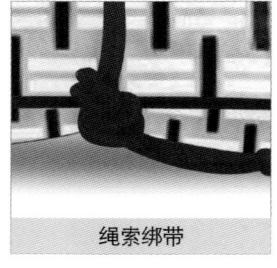

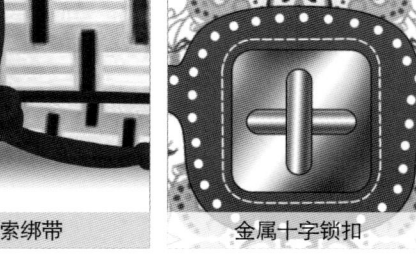

绳索绑带　金属十字锁扣　牛皮压花

皮质流苏

淡色的男式手拿包，简洁的款式，配上欧式花纹，具有简洁从容新贵感。
Elegant prints add noble sensibilities to this light color men's clutch featuring a sleek silhouette.

皮革雕花

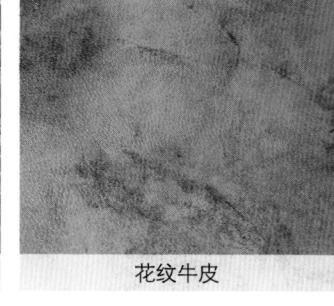

花纹牛皮

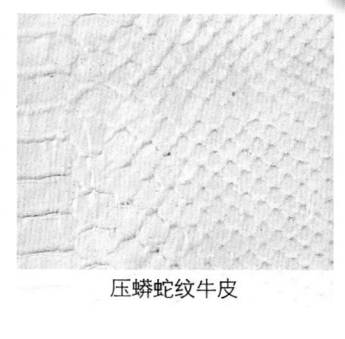

压蟒蛇纹牛皮

欧式印花牛皮

细密荔枝纹牛皮

S 箱包 2015 春夏海派时尚流行趋势

海上星梦
STAR DREAMS OVER THE SEA

圣诞节外滩装饰

静安寺夜景

路边咖啡厅

欧式风格灯饰

花卉图案餐具

复古胭脂盒

室内灯饰摆i

设计灵感

在拉杆箱上用压花牛皮或印花面料作为主要面料，搭配有质感的皮扣、包边、包角把手等进行对比。

The hard shell trip packing case features leather hemming and prints decoration. The leather one features elegant prints.

鸵鸟皮

压花牛皮

磨砂牛皮

80　www.style.sh.cn

2015 SPRING/SUMMER STYLE SHANGHAI FASHION TREND / BAG&SUITCASE

关键要素 KEY POINTS | 新贵气质 NOBLE TEMPERAMENT

曜石黑

迷彩棕

剑麻灰

浪花白

岩石灰

包面凹凸压花的双肩包，配合同色系深色的手缝装饰。
Dark tone chains and a plush quilted panel further the vibe of a roomy backpack.

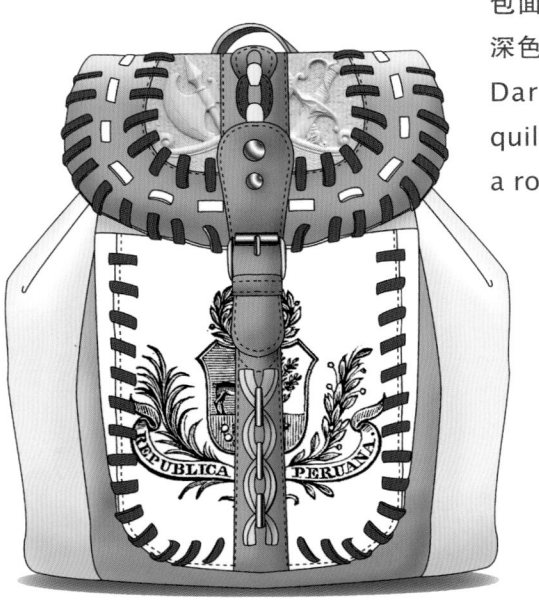

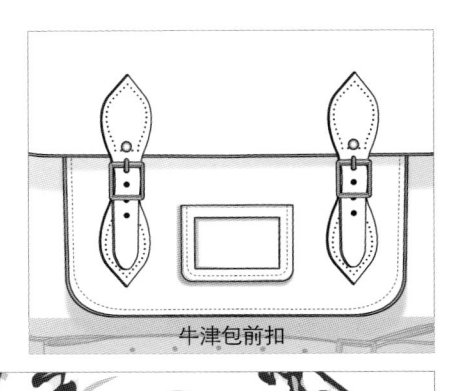
牛津包前扣

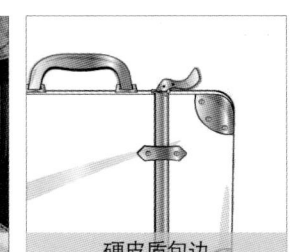
粗手缝扣

英式边条花纹及缝线

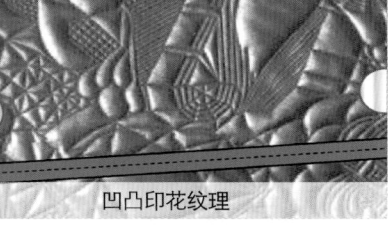
凹凸印花纹理

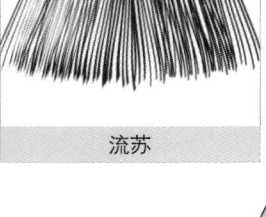
流苏

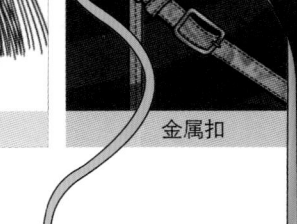
金属扣

硬皮质包边

五金锁扣配件

流苏圆形镂空童包，半圆形的镂空皮质设计感，配合大面积的长流苏，具有灵动活泼感。
Swishy tassels add vitality to this round shaped kid bag with engraving leather.

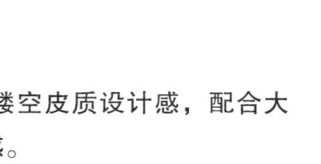

亮面磨砂牛皮

猪皮

欧式印花布

荔枝纹牛皮

大理石纹牛皮

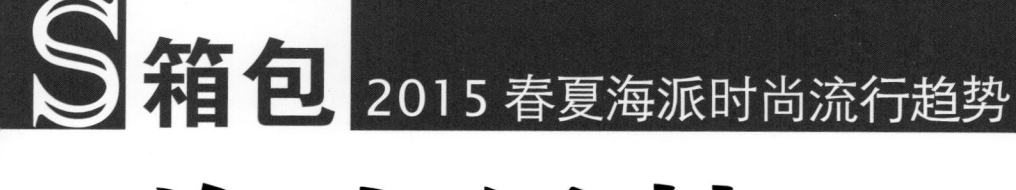

海上星梦
STAR DREAMS OVER THE SEA

创新视觉表达方式 　创新材料装饰 　新材料灯具展示 　宜家灯饰

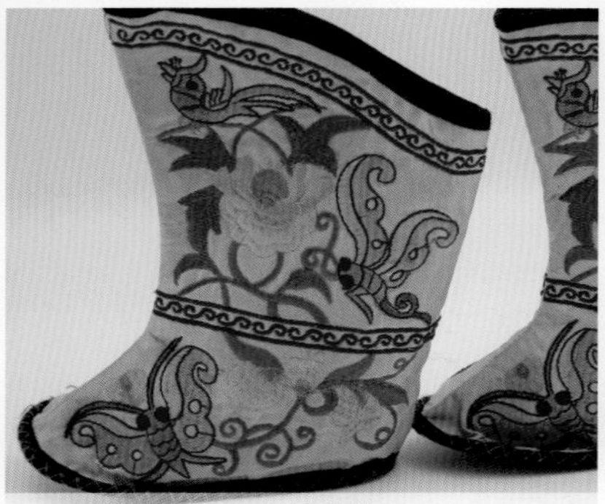

古代绣花鞋 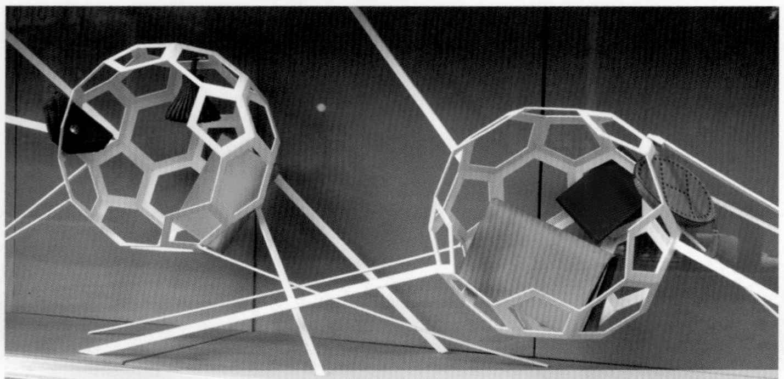　橱窗布置 　用布艺手法仿土

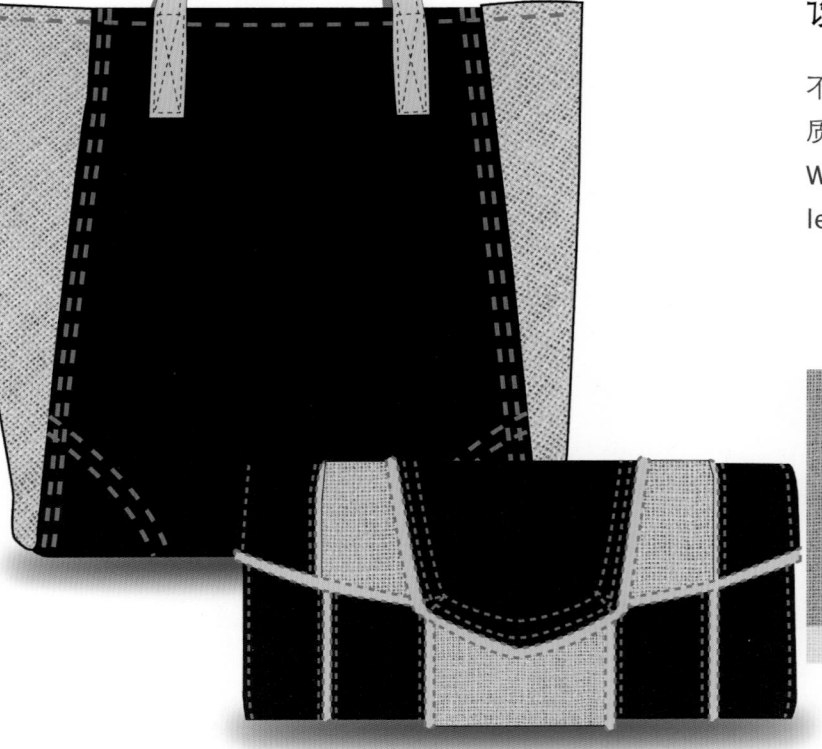

设计灵感

不同材料不同颜色的拼接是创新包的主旨，麻质材料、牛仔面料和牛皮材质的拼接，配上粗的不同颜色的手缝线的对比，在视觉上有创新的感觉。
We create this innovative bag using linen together with denim and leather.

麻质材料

牛仔材料

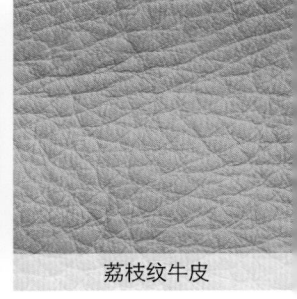
荔枝纹牛皮

82　www.style.sh.cn

2015 SPRING/SUMMER STYLE SHANGHAI FASHION TREND / **BAG&SUITCASE**

关键要素 | 创新传统
KEY POINTS | INNOVATED TRADITIONS

淡奶黄

皮革黄

斗笠黄

紫烟蓝

青墨蓝

透明 PVC 面料配上牛仔布料，皮质的包边。
Transparent PVC with denim patches
and leather edge covering.

包肩带

粗手缝线

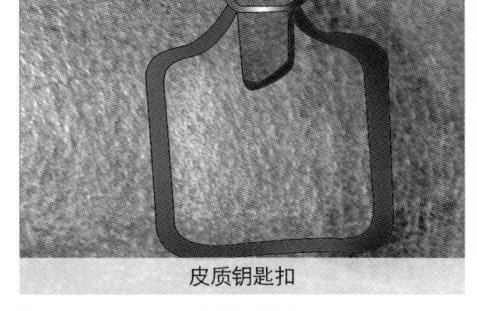
皮质钥匙扣

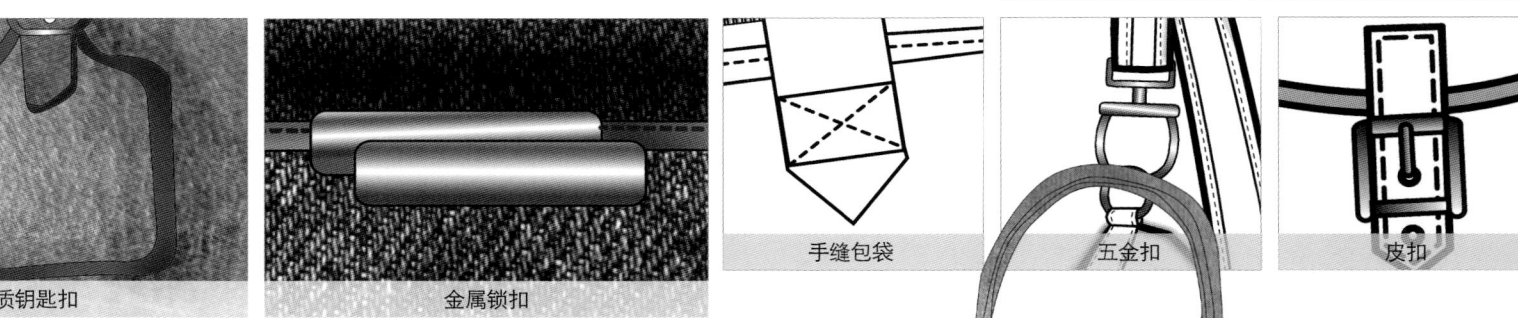
金属锁扣

手缝包袋

五金扣

皮扣

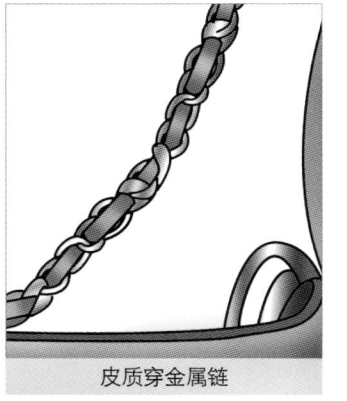
皮质穿金属链

皮质和牛仔布拼接的包袋，口金样式的开口，配上皮质的搭扣。
Denim patches and leather buckle add vitality
to this metal hardware topped leather bag.

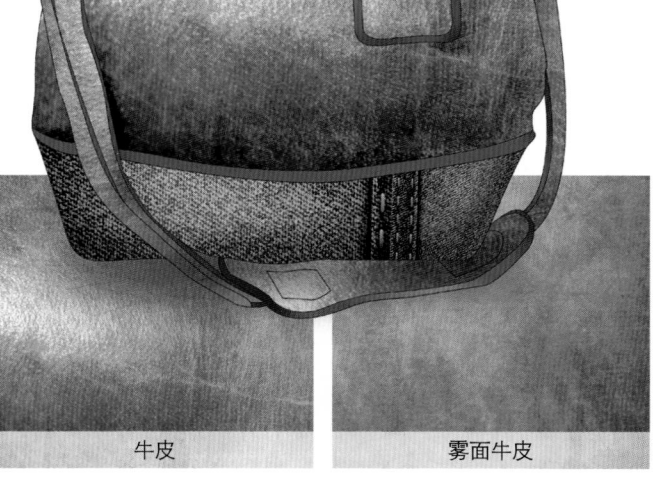

透明 PVC

粗牛仔布料

做旧牛仔布料

牛皮

雾面牛皮

83

S箱包

2015 春夏海派时尚流行趋势

优活之梦
LIVING THE DREAM

优活之梦箱包多采用生态材质和自然元素，平民贵族箱包以优雅的轮廓结合舒适的面料彰显低调中的奢华。家庭手工艺则采用精致的细节塑造箱包，节俭设计以拼接为主要特色，生态材质质朴的设计风格让你瞬间释放压力，尽情感受大自然带给你的美好体验。

"LIVING THE DREAM" collection features eco-materials and natural elements while "ORDINARY NOBILITY" combines elegant silhouette with comfortable fabrics, low-key but luxurious. "FAMILY HANDICRAFTS" has very exquisite details while "SIMPLE DESIGN" uses stitching a lot. "ECO MATERIAL" will give you a relaxed feeling with rustic design and let you enjoy what nature brings us.

BAG&SUITCASE

2015 SPRING/SUMMER STYLE SHANGHAI FASHION TREND
海派箱包流行趋势

2015 SPRING/SUMMER STYLE SHANGHAI FASHION TREND / **BAG&SUITCASE**

关键要素 | KEY POINTS

织补风尚
BACK TO THE BASICS

平民贵族
ORDINARY NOBILITY

生态材质
ECO MATERIAL

家庭手工
FAMILY HANDICRAFTS

淡奶黄

抹茶绿

苔藓绿

曜石黑

青花蓝

孔雀蓝

水晶蓝

豆浆白

香芋灰

紫砂棕

S 箱包 2015 春夏海派时尚流行趋势

优活之梦
LIVING THE DREAM

旧储物盒改造成的钟

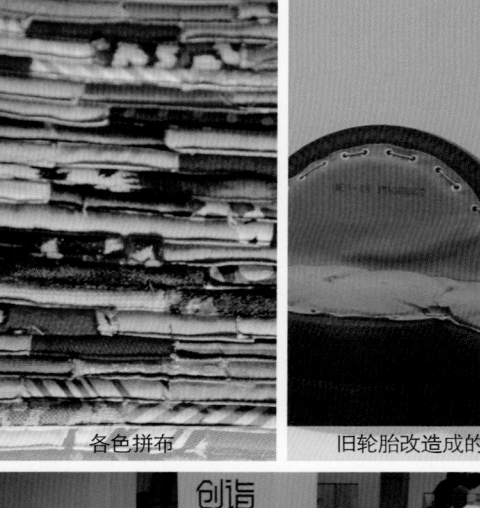
各色拼布

旧轮胎改造成的沙发

破损状麻布

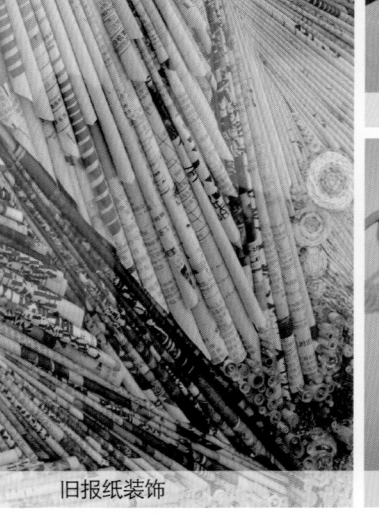
旧报纸装饰

旧电话改造成的钟

旧杂志报纸陈列设计

设计灵感

运动包采用麻质面料，给人一种利落干净的感觉。双麻布拼接打破了沉闷而凸显活力。做旧破洞点缀轻松自然又充满岁月感和质感。
Tough canvas ensures durability in a versatile duffel bag decorated with destroyed holes for a rustic feel.

粗麻布

花样麻布

草纤维面料

86   www.style.sh.cn

2015 SPRING/SUMMER STYLE SHANGHAI FASHION TREND / BAG&SUITCASE

关键要素 | 织补风尚
KEY POINTS | BACK TO THE BASICS

拼接面料搭配竹编材质让人有种回到20世纪的感觉。简洁的包型，大方、轻松、自然。

A perfectly proportioned bag in sleek materials is embellished with braided straps.

将环保袋和手提包元素相结合设计。无论外在和内在都体现出节俭、环保的概念。

Add eco bag elements to women handbag, to express the eco concept from both the shape and structure.

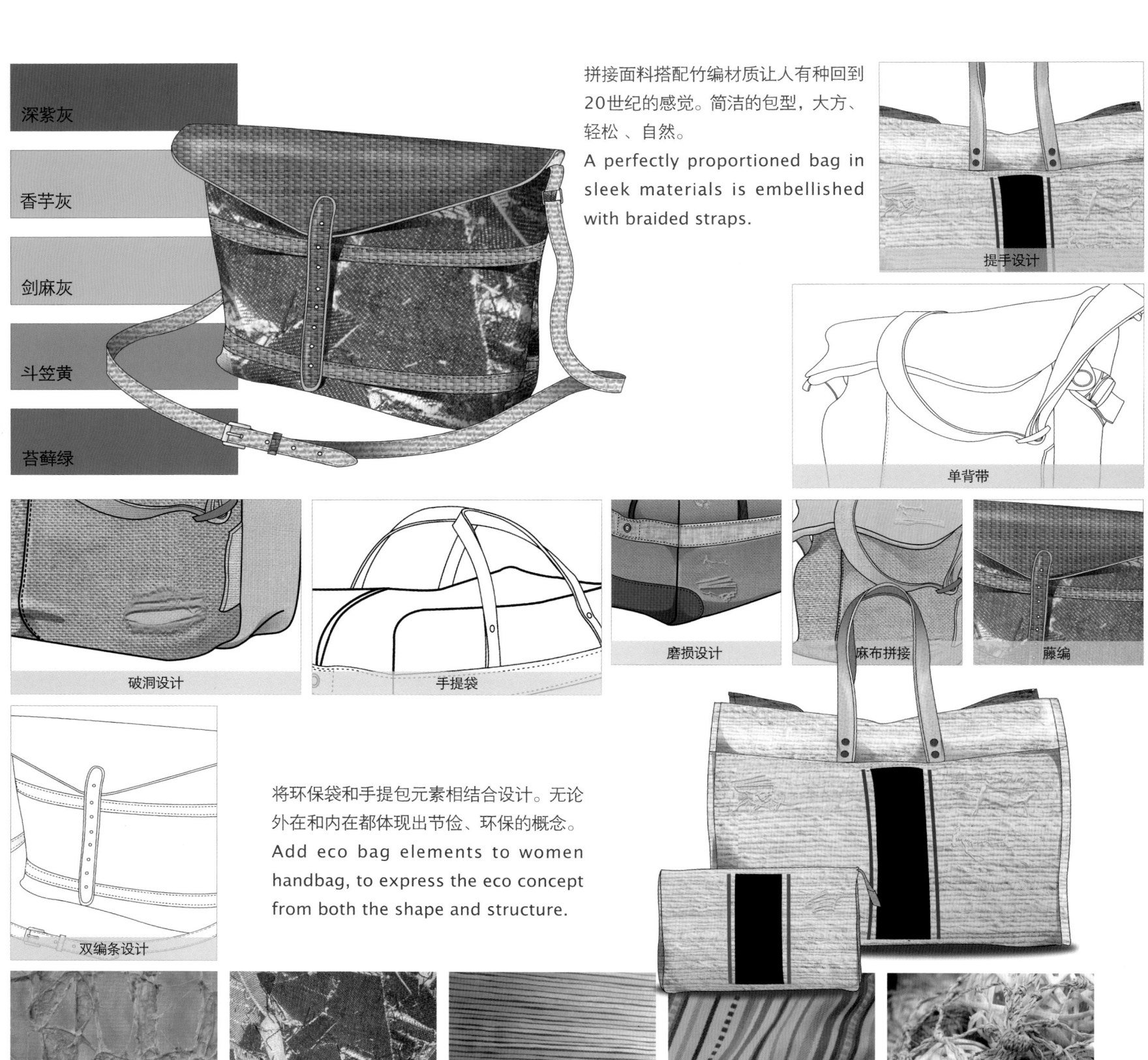

深紫灰

香芋灰

剑麻灰

斗笠黄

苔藓绿

提手设计

单背带

破洞设计

手提袋

磨损设计

麻布拼接

藤编

双编条设计

做旧面料

拼布面料

麻布

印花布料

麻绳

S 箱包 2015 春夏海派时尚流行趋势

优活之梦
LIVING THE DREAM

旧热水瓶改造成的灯具

旧电视机与电子产品结合

老糖罐改造成的灯

建筑外观木质材料的运用

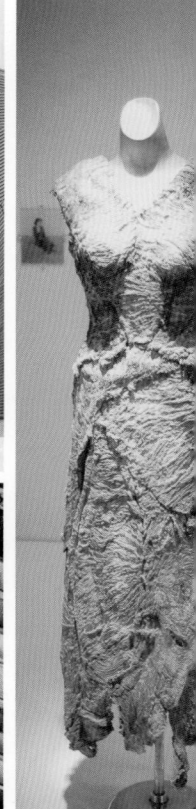
废材料服装

纸质鞋子

各色旧麻布拼接

设计灵感

童包采用了拼布和做旧设计。较高明度的印花图案更衬托小朋友们的纯洁和天真。包型设计圆润饱满，深受小朋友们的喜爱。

Python highlights the cuteness and quality of this round shaped kids' bag.

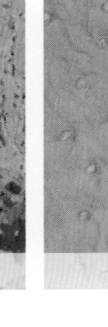
软木印花花纹面料

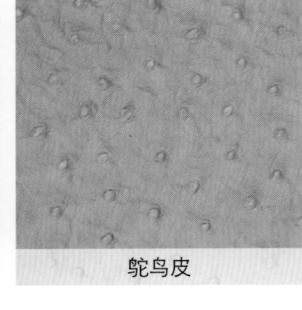
鸵鸟皮

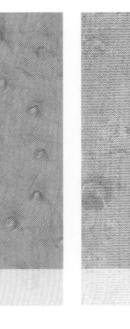
印花麻布

88　www.style.sh.cn

2015 SPRING/SUMMER STYLE SHANGHAI FASHION TREND / BAG&SUITCASE

关键要素 | 织补风尚
KEY POINTS | BACK TO THE BASICS

相对于男包和运动包而言。女包和童包加入了淡雅的印花作为装饰。面料上仍然延续男包和运动包的风格。
Light colored prints add feminine elements to women's and kids' handbags. Materials used here are very common in men's bag and backpacks.

淡雅的花卉图案与麻布结合,体现自然、节俭的同时,也展现了女性的妩媚和柔情。软木和麻布的结合也很清新。
Elegance floral prints on linen cloth look so natural and feminine.

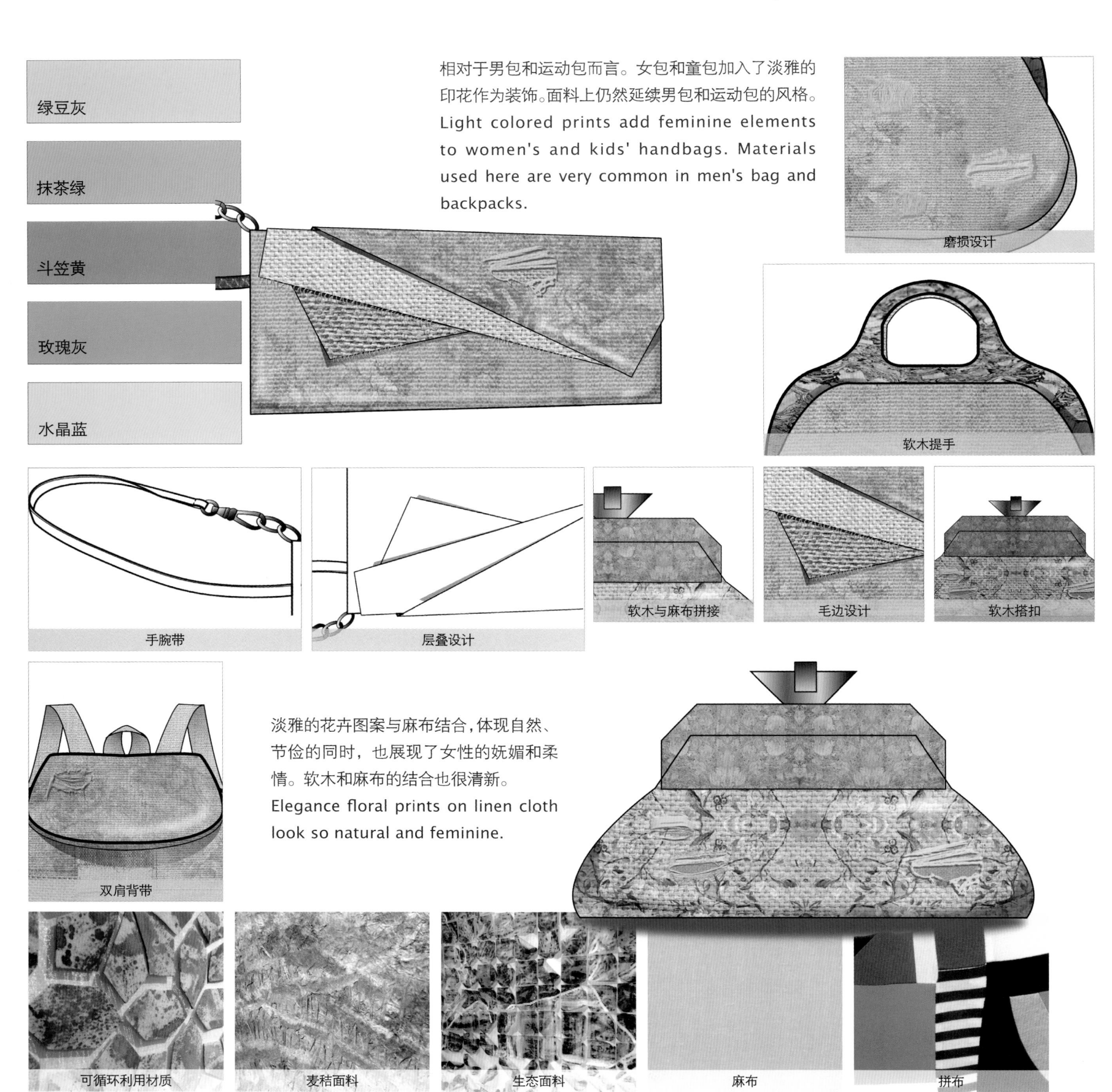

绿豆灰
抹茶绿
斗笠黄
玫瑰灰
水晶蓝

磨损设计

软木提手

手腕带

层叠设计

软木与麻布拼接

毛边设计

软木搭扣

双肩背带

可循环利用材质

麦秸面料

生态面料

麻布

拼布

S箱包

2015 春夏海派时尚流行趋势

优活之梦
LIVING THE DREAM

箱包雕塑

平民饰品

平民生活用品

纸质服装

贵族化妆用品

简洁沙发

雕花墙面

设计灵感

手提女包体现女性的优雅知性。包袋上的人造皮采用鳄鱼纹压花。平民面料也能制造贵族美感。淡雅的色调让包袋更经典耐看。

A classic handbag styled with exposed special tree-bark texture is quite unique.

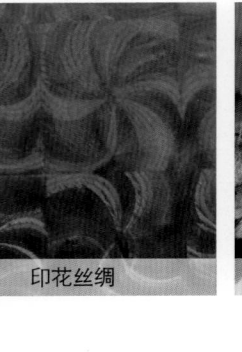

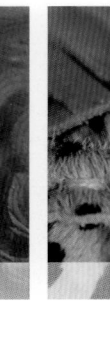

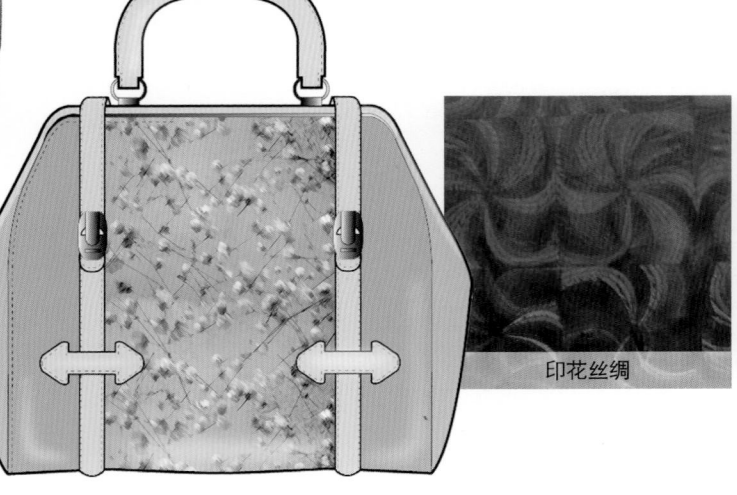

印花丝绸

动物纹皮料

人造鳄鱼纹皮

90 www.style.sh.cn

2015 SPRING/SUMMER STYLE SHANGHAI FASHION TREND / BAG&SUITCASE

关键要素 | 平民贵族
KEY POINTS | ORDINARY NOBILITY

荔枝棕

绿豆灰

剑麻灰

香芋灰

豆浆白

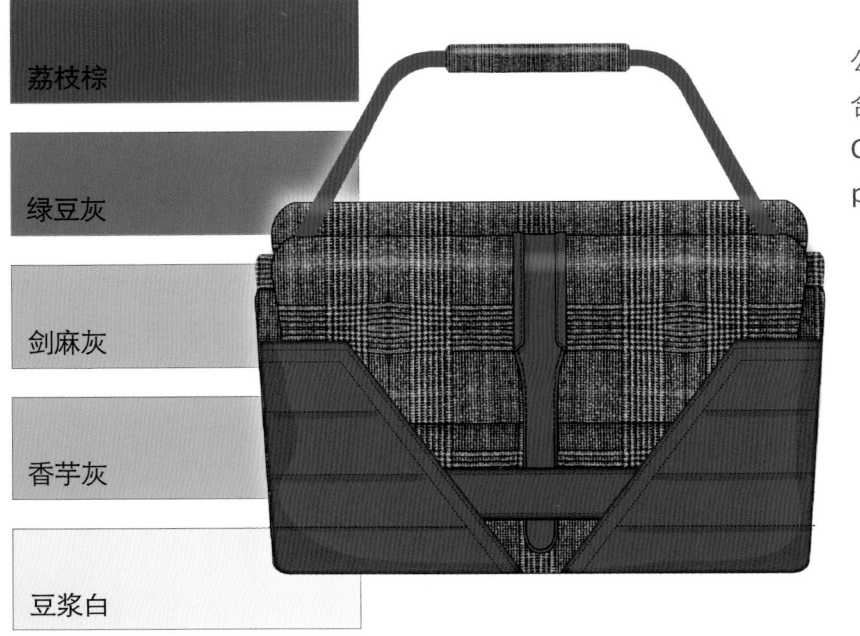

公文包款式简洁大方。皮料与麻布混合，彰显贵气的同时又有亲民感。
Classic briefcase with new floral prints.

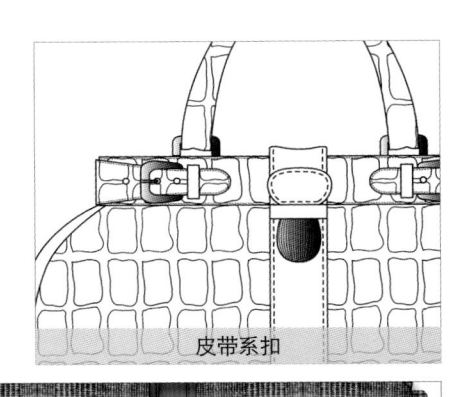

皮带系扣

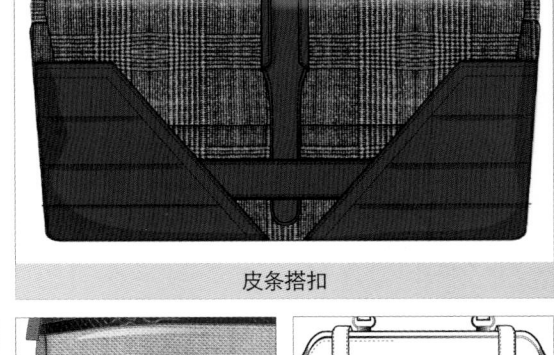

皮条搭扣

麻布与皮拼接对比

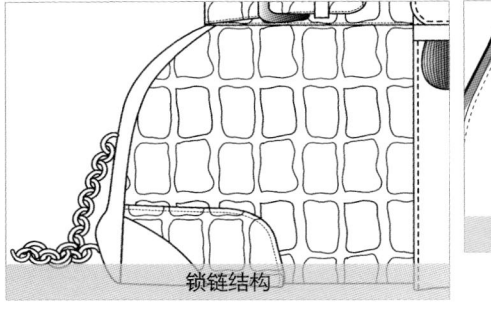

锁链结构

背带搭扣

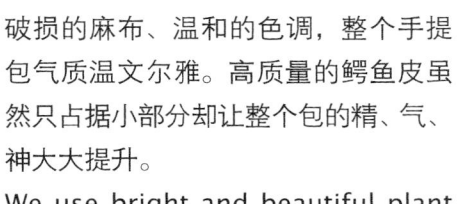

磨损设计

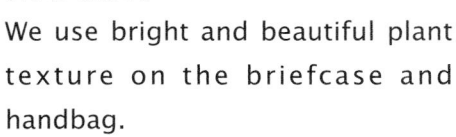

皮背带条

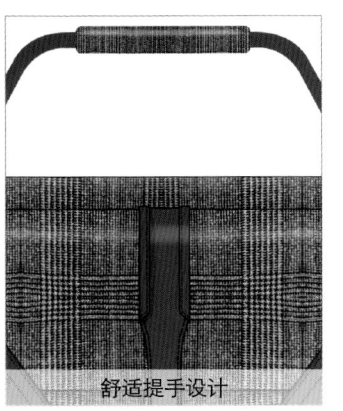

舒适提手设计

破损的麻布、温和的色调，整个手提包气质温文尔雅。高质量的鳄鱼皮虽然只占据小部分却让整个包的精、气、神大大提升。
We use bright and beautiful plant texture on the briefcase and handbag.

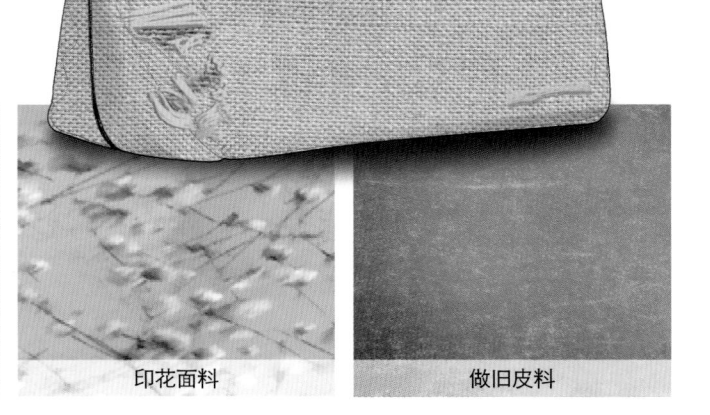

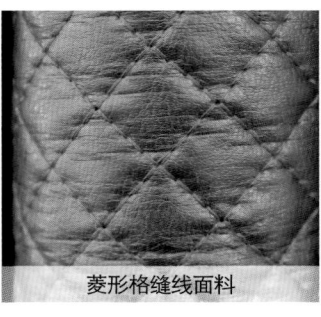

菱形格缝线面料

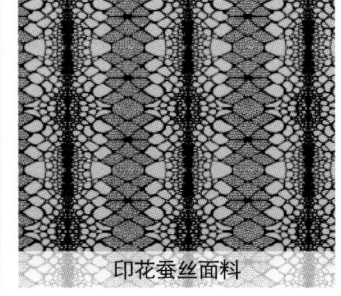

印花蚕丝面料

粗麻布

印花面料

做旧皮料

91

S 箱包 2015 春夏海派时尚流行趋势

优活之梦
LIVING THE DREAM

平民手工艺品

华丽的吊顶

农家一角

廉价精致烛台

精致陶瓷茶具

破旧的房墙

精致衬衫

设计灵感

童包款式设计运用圆润的轮廓，体现童包的可爱与活泼。面料讲究舒适性，注重对儿童的关怀。鳄鱼皮的点缀使包袋显得更有精神。
Python highlights the cuteness and quality of this round shaped kids' bag.

素布底蕾丝   粗麻布   荔枝纹牛皮

92　www.style.sh.cn

2015 SPRING/SUMMER STYLE SHANGHAI FASHION TREND / BAG&SUITCASE

关键要素 | 平民贵族
KEY POINTS | ORDINARY NOBILITY

平民与贵族，看似矛盾却又充满对比趣味。这个主题的箱包采用麻布和皮料的对比为主要特色。
This collection creatively features linen materials with contrasting leather.

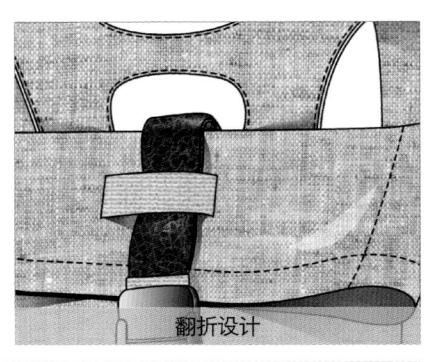

翻折设计

紫砂棕

枣泥红

斗笠黄

豆沙灰

蛋壳黄

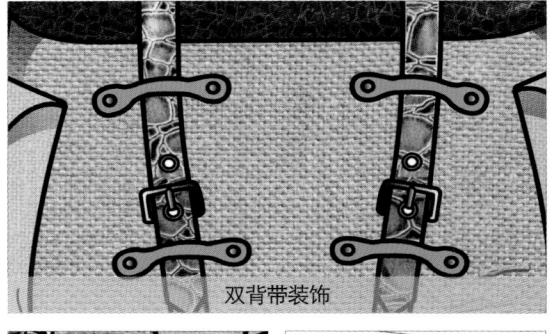

双背带装饰

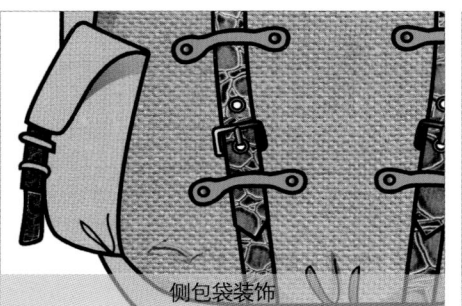
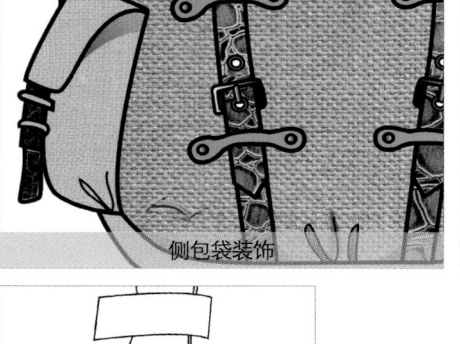

侧包袋装饰

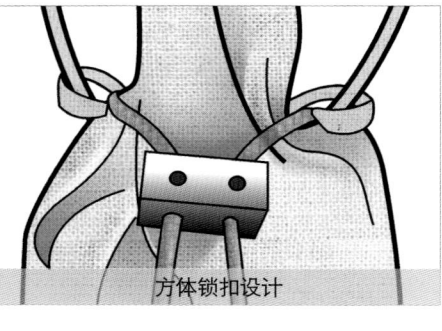

方体锁扣设计

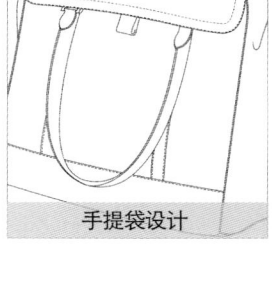

手提袋设计

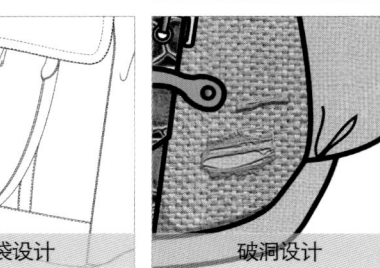

破洞设计

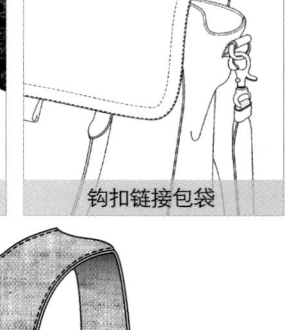

钩扣链接包袋

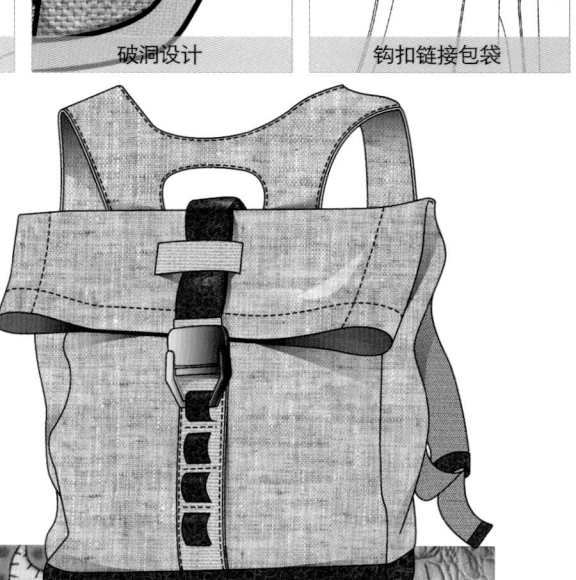

运动包运用透气的麻布为主要面料，给人一种轻盈爽朗的感觉，点缀的皮料又营造出贵族的气息。
We use breathable linen in this sports bag and decorate with some straps of leather.

搭扣设计

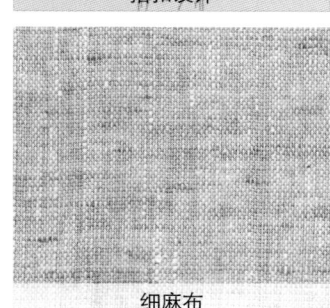

细麻布

蛇皮纹压花皮料

牛皮

印花布料

鳄鱼皮

93

S箱包 2015 春夏海派时尚流行趋势

优活之梦
LIVING THE DREAM

海星与花卉

密集的绿叶

仿生灯饰

猪笼草

自然花卉纹样

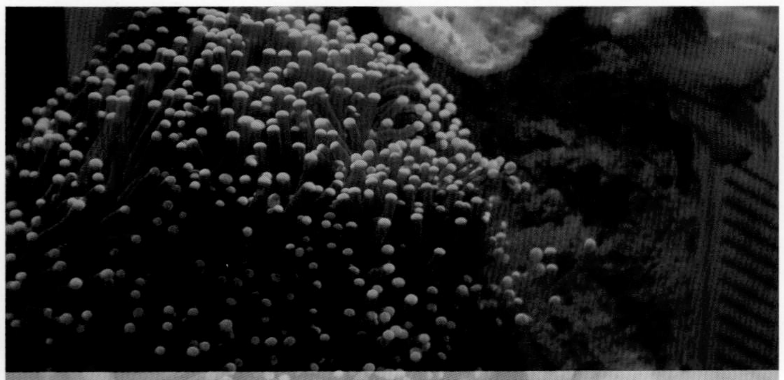

海洋生物

茂盛的绿叶

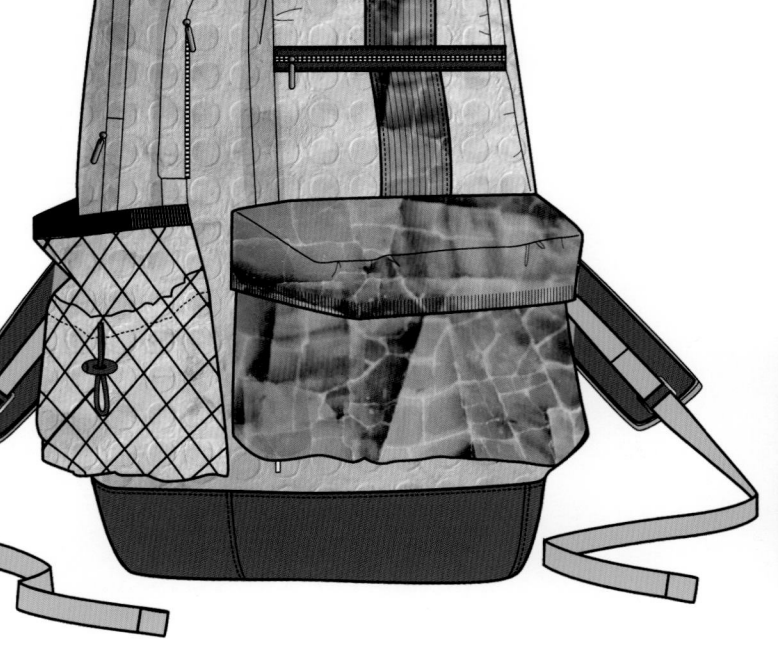

设计灵感

独特的材质给多功能运动包的设计注入了新鲜的血液，给人以新鲜感。仿叶脉肌理面料仿佛让背包能呼吸。

Leaf vein texture makes this functional sports bag very fresh looking. It seems that it can breathe as a real leave.

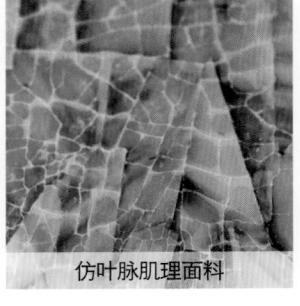

仿叶脉肌理面料

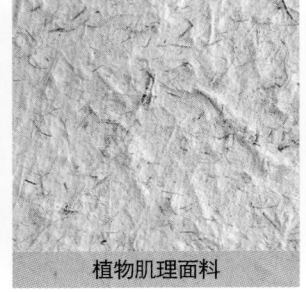

植物肌理面料

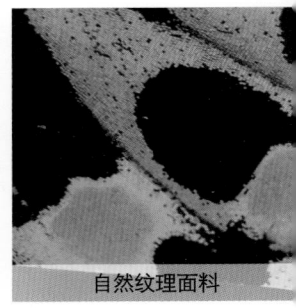

自然纹理面料

94   www.style.sh.cn

2015 SPRING/SUMMER STYLE SHANGHAI FASHION TREND / **BAG&SUITCASE**

关 键 要 素 | 生态材质
KEY POINTS | ECO MATERIAL

青花蓝

孔雀蓝

紫砂棕

抹茶绿

豆浆白

大地色系的腰包搭配生态材质让人仿若置身于一个奇幻的森林中。
Earth tone waist bag exudes natural beauty with eco-friendly material.

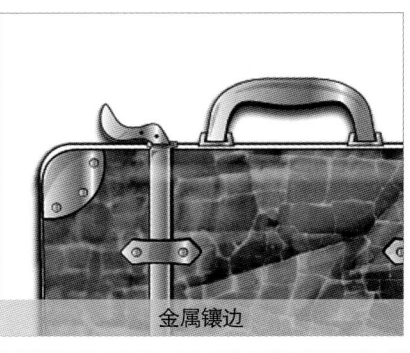
金属镶边

层叠设计

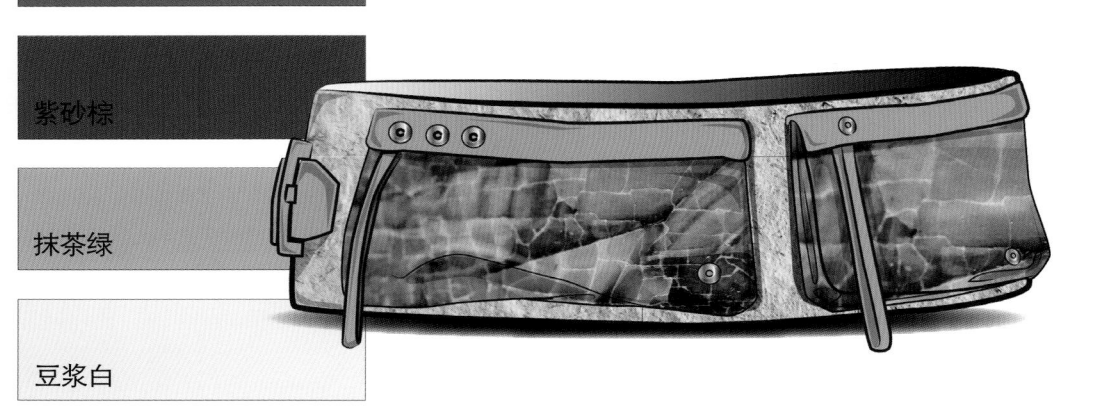

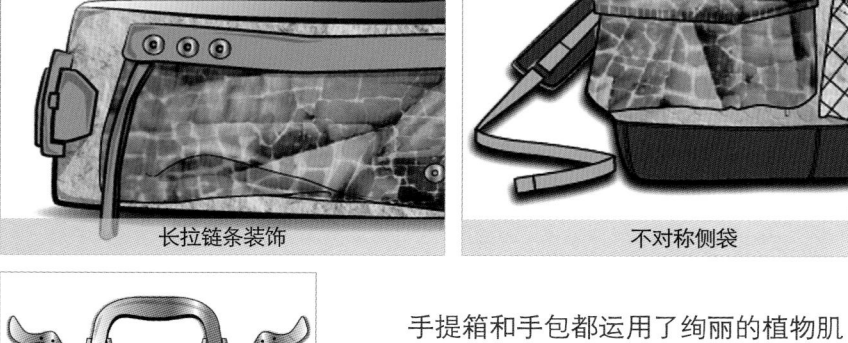
长拉链条装饰

不对称侧袋

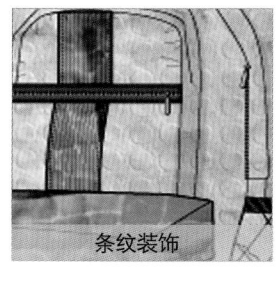
条纹装饰

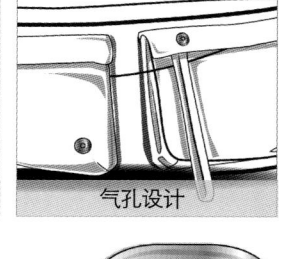
气孔设计

松紧网袋

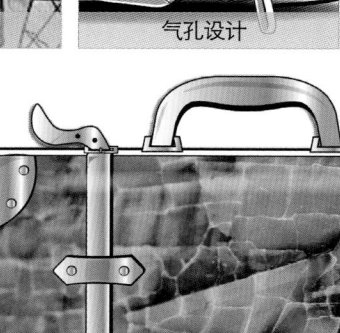
背带设计

手提箱和手包都运用了绚丽的植物肌理面料，充满自然气息。
We use bright and beautiful plant texture on the briefcase and handbag.

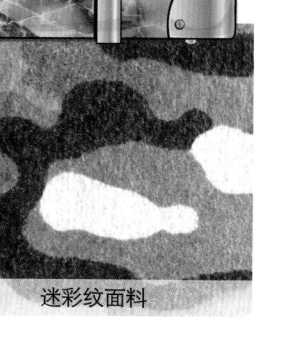

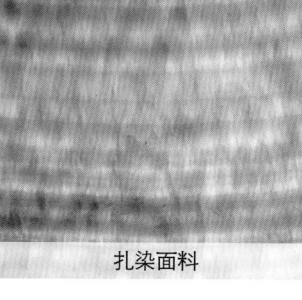
扎染面料

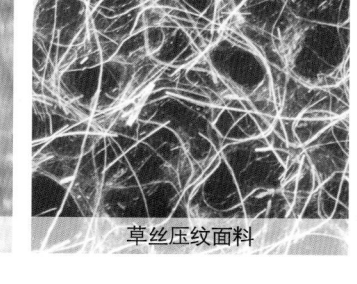
草丝压纹面料

透明树脂

有机植物面料

迷彩纹面料

95

S 箱包 2015 春夏海派时尚流行趋势

优活之梦
LIVING THE DREAM

竹编风格建筑

海派建筑纸雕

传统钩编作品

竹编器皿

刺绣线

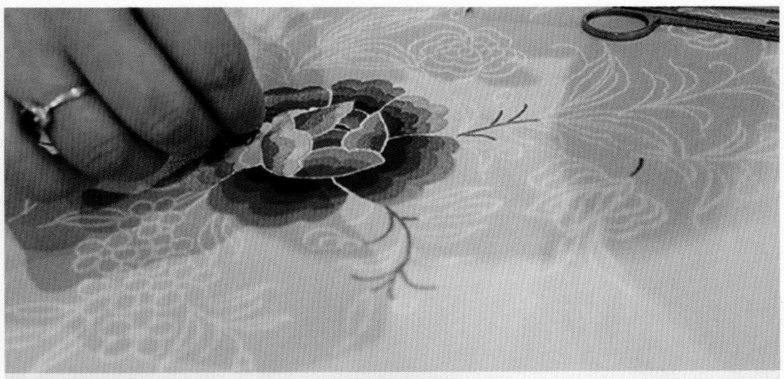

传统刺绣图案

手工艺术品

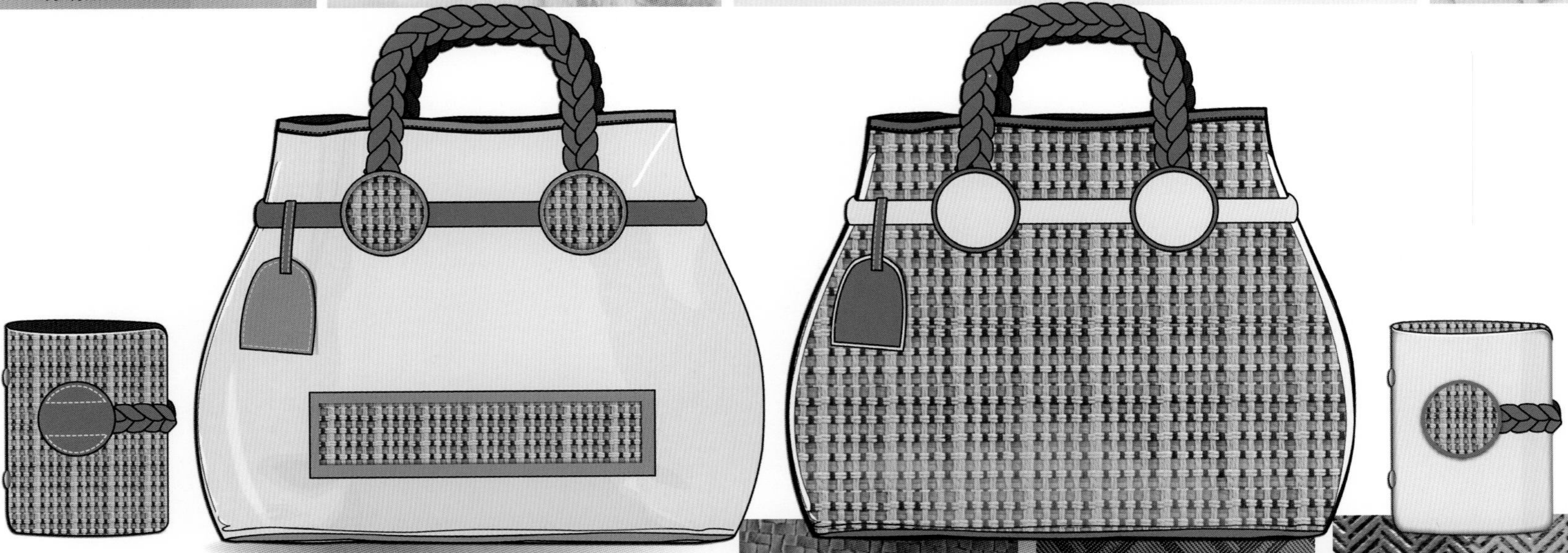

设计说明

简洁包型采用编织面料，减弱其他设计元素，突出手工编织的原汁原味，钱包也运用编织元素显得小巧精致。

A woven design in a classic and cute wallet.

编制皮革

竹编纹路纹样

编织方式镂空面料

96   www.style.sh.cn

2015 SPRING/SUMMER STYLE SHANGHAI FASHION TREND / BAG&SUITCASE

关键要素 | 家庭手工
KEY POINTS | FAMILY HANDICRAFTS

皮料编织面料起到低调沉稳的装饰效果。做旧牛皮也彰显了低调的华丽。
Richly woven leather keeps you stylishly organized.

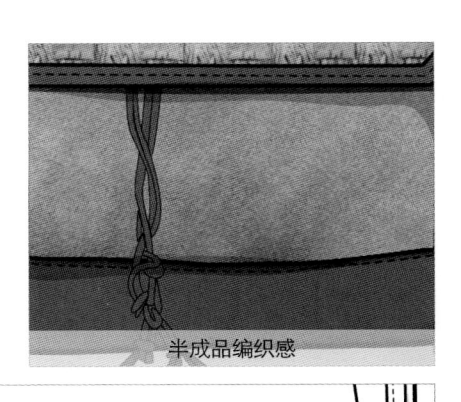

半成品编织感

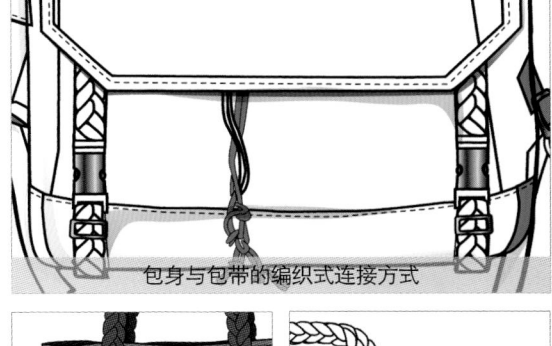

包身与包带的编织式连接方式

紫砂棕

苔藓绿

斗笠黄

蜜瓜黄

艾草绿

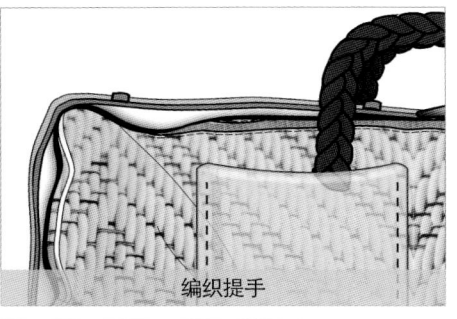

编织提手

钩编装饰挂坠

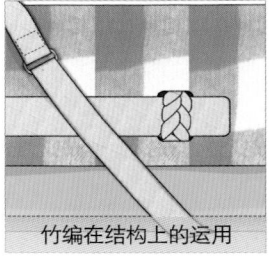

竹编在结构上的运用

花式钩编包带　手工钩编包身

彩色藤编面料表面加上透明 PVC 面料显得潮味十足。独特的编织提手让整个包显得独具特色。
Colorful woven design in a transparent PVC handbag makes it stylish.

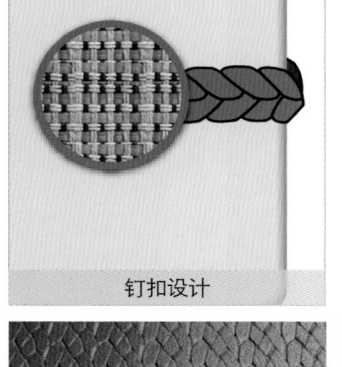

钉扣设计

钩编皮革

竹编结构复合面料

手工艺花卉图案

多彩创新编织

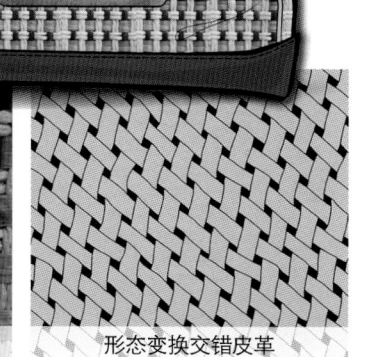

形态变换交错皮革

97

S 箱包 2015 春夏海派时尚流行趋势

时空筑梦
REALISING DREAMS THROUGH SPACE

以淮海路 K11 为例的时尚新地标正在上海逐渐崛起，其内部装饰前卫大胆，并与人文、自然相结合，以独特的姿态屹立于上海这个以"光速"发展的城市，并形成了自己的独特魅力。

New fashion landmarks such as K11 are emerging in Shanghai. The modern interior design combines humanism with nature creating a unique and charming shopping mall.

BAG&SUITCASE

2015 SPRING/SUMMER STYLE SHANGHAI FASHION TREND
海派箱包流行趋势

2015 SPRING/SUMMER STYLE SHANGHAI FASHION TREND / BAG&SUITCASE

关键要素 | KEY POINTS

复制粘贴
REPETITION & MODULIZATION

符号人生
SYMBOLIC LIFE

机械生物
MECHANICAL CREATURE

单纯设计
SIMPLE DESIGN

珍珠白

氯化灰

碳钢灰

天光蓝

光影蓝

亚铜灰

陨石棕

烟云灰

沙砾黄

苹果绿

S箱包 2015 春夏海派时尚流行趋势

时空筑梦
REALISING DREAMS THROUGH SPACE

薄墙立柱

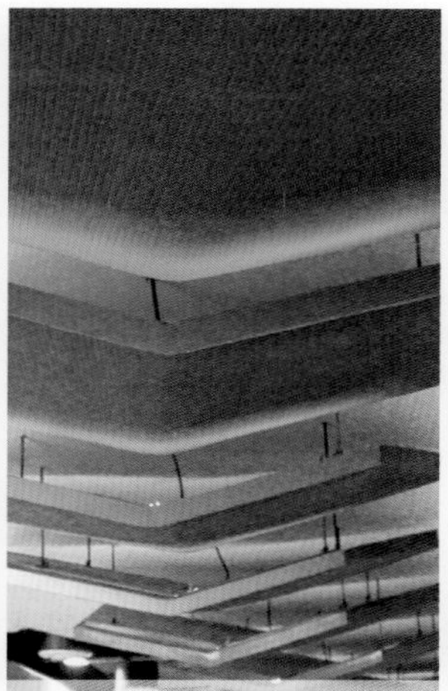

三角屋顶

图形重复

小蜂巢大蜂巢

平面三维

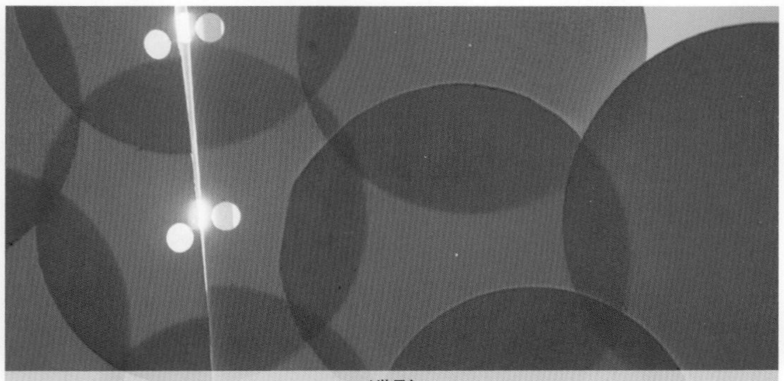

谍影

圆柱感

设计灵感

三维立体的复制粘贴表现在包上的元素就是图形的视错觉。星星点点的波点、线条和仿立体块面营造女性包袋的多样化。

Repetition of stars, dots, lines and blocks creates 3D illusion on bags.

条纹棉布

纸质面料

针织面料

100   www.style.sh.cn

2015 SPRING/SUMMER STYLE SHANGHAI FASHION TREND / **BAG&SUITCASE**

关 键 要 素 | 复 制 粘 贴
KEY POINTS | REPETITION & MODULIZATION

光影蓝

天光蓝

日光白

烟云灰

碳钢灰

延伸的条带设计突破了传统，视觉冲击感强，适合整体简单的装束。

Extended long strap may attract more attention due to visual impact.

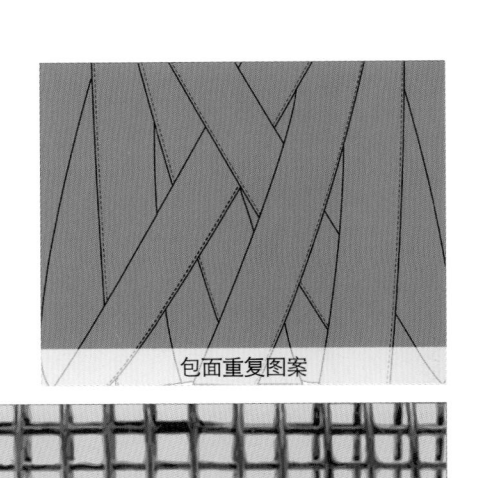

包面重复图案

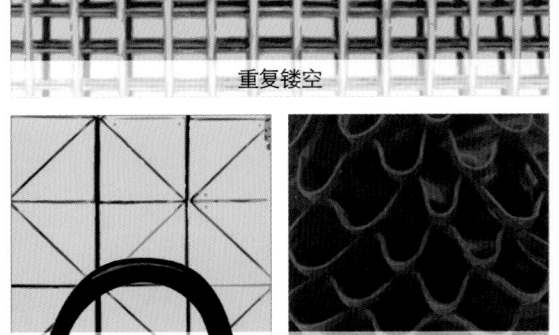

重复镂空

重复图形

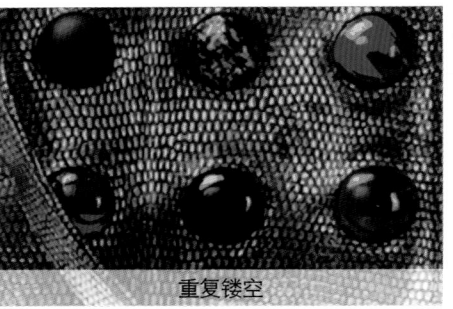

重复镂空

重复装饰

重复图形

重复镂空

重复装饰

这款黑白组合的设计点在于线条分割，包面上的重复块面组合使原本简单的设计变得灵动。

Repetition of blocks on this black and white bag makes it look more dynamic.

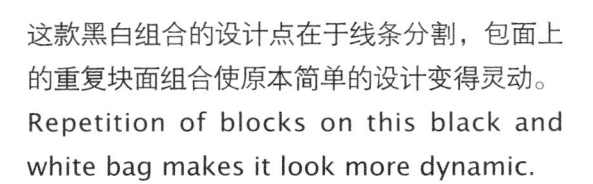

立体印花面料

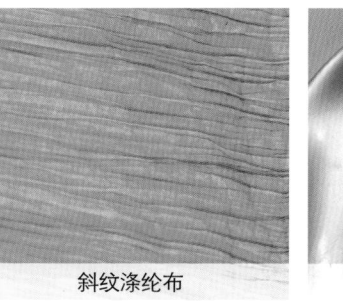

斜纹涤纶布

银光色皮料

牛皮

混纺面料

101

 **箱包** 2015 春夏海派时尚流行趋势

时空筑梦
REALISING DREAMS THROUGH SPACE

重复的阶梯分割线

重复变化的屋顶骨骼

图形重复

立体构成作品中重复的骨骼

灯罩的重复

灯具中重复透叠的视觉效果

重复的立体及线

设计灵感

未来复制在包袋上的表现整体较素净，结构的重复堆叠是主打元素，镂空的外部可透出迷离的内部结构。整体上多了点似有似无，面料以立体材质为主。

We can look through the engraving exterior into fancy interior. Repetition of basic structures is the feature of this piece of futuristic design.

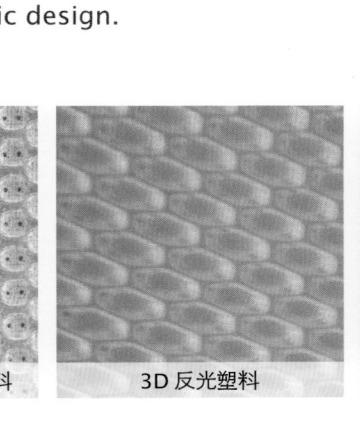

带有几何图案的针织面料

3D 反光塑料

透明 PVC

102   www.style.sh.cn

2015 SPRING/SUMMER STYLE SHANGHAI FASHION TREND / **BAG&SUITCASE**

关 键 要 素 | 复制粘贴
KEY POINTS | REPETITION & MODULIZATION

氯化灰

珍珠白

碳钢灰

日光白

岩石灰

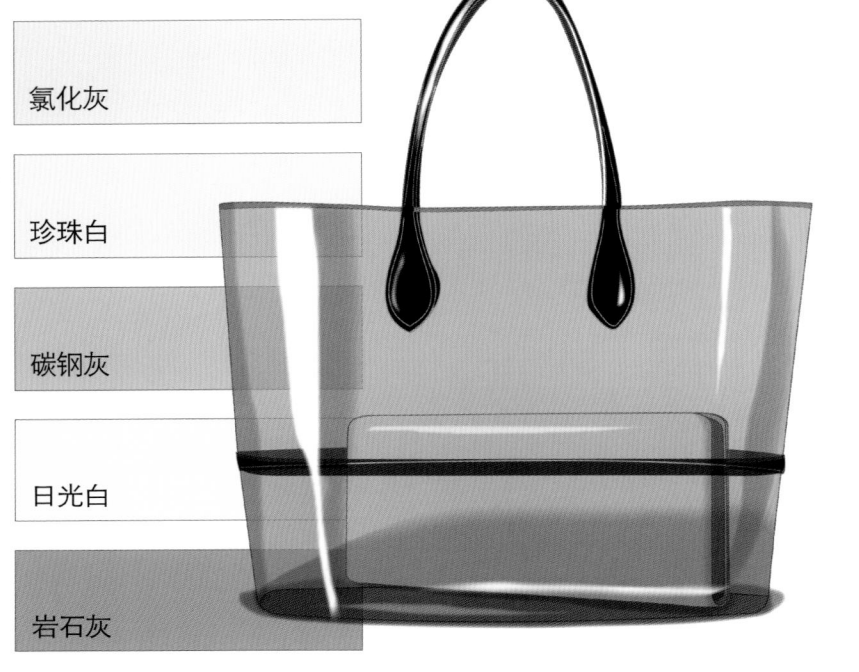

女包中的一个亮点是轮廓的复制，先有一个大的外轮廓，内部有一个可以随意更换的小轮廓，构成了复制的空间效果。

An innovation in women's handbag is a 2-piece-in-1-set design. First we pick a sleek silhouette then add in a changeable small part.

包面贴图聚拢复制

内部透叠

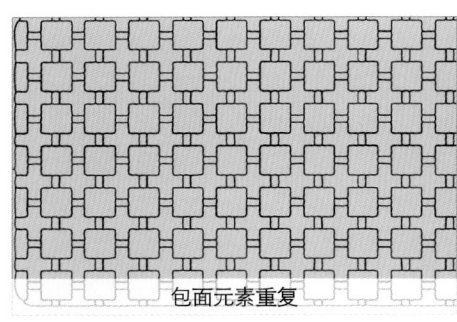

包面元素重复

包面重复镂空

包面立体复制

包面凹体复制

包面金属条纹复制

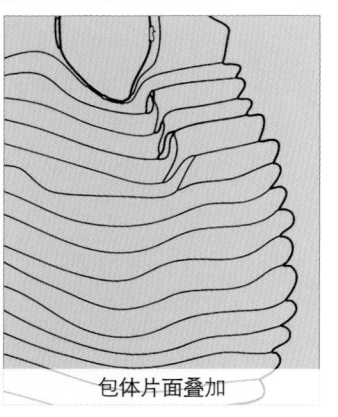

包体片面叠加

此款女包外轮廓的处理值得借鉴，仿佛压花亚克力的外壳是其设计点。内部可以根据喜好更换任意包包，实用性极强。

Application of 2-piece-in-1-set design.

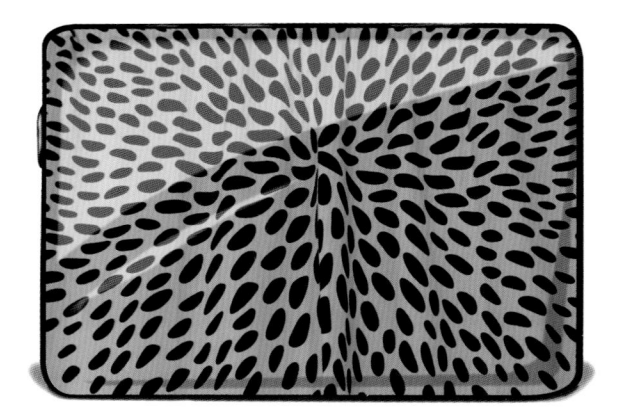

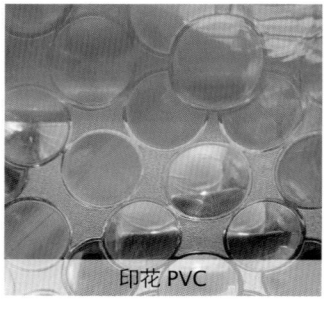

印花 PVC

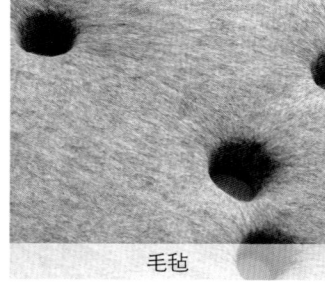

毛毡

印花牛津布

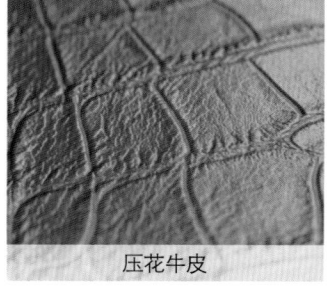

压花牛皮

镂空编织牛皮

103

S箱包 2015 春夏海派时尚流行趋势

时空筑梦
REALISING DREAMS THROUGH SPACE

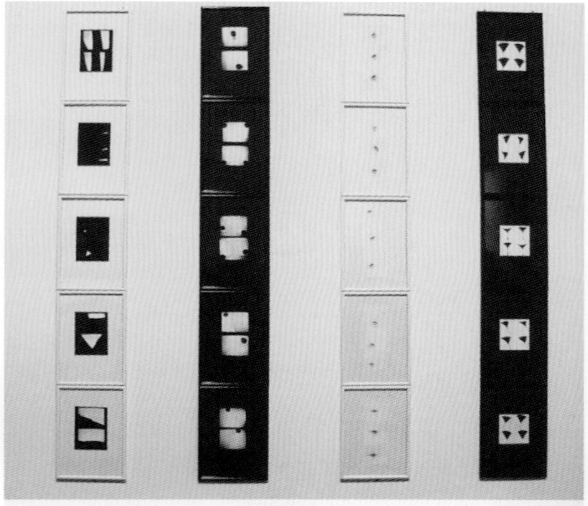

相片的新组合

新极简主义

有故事的图画

背光墙壁

灯光亮起

想象里面是什么

地铁图再设计

设计灵感

可以从动物中寻找灵感，观察动物的生活习惯，提炼动物的肢体语言，来表现人类的设计构思，并会成为下一季的亮点。

Try to find inspiration from animals. Observing how animals act and communicate and trying to transform it into design language are pretty inspiring.

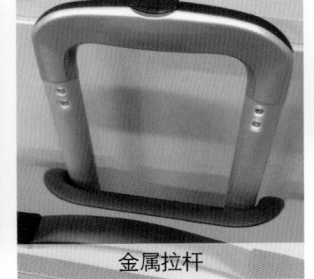

PVC 材质

金属拉杆

塑料万向轮

104   www.style.sh.cn

2015 SPRING/SUMMER STYLE SHANGHAI FASHION TREND / BAG&SUITCASE

关键要素 | 符号人生
KEY POINTS | SYMBOLIC LIFE

亚铜灰

天光蓝

日光白

氯化灰

碳钢灰

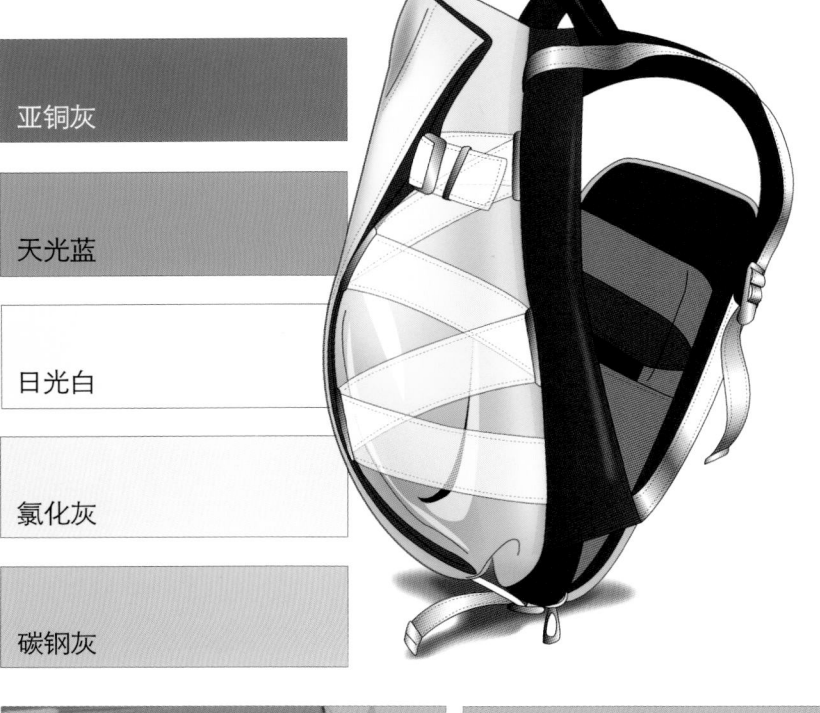

设计灵感来源于船的造型，双开门设计可以放置专属的贵重物品，背上后从后面看形似江中小船。

This bag is inspired by boat. Two pockets can be used to put valuable personal belongings separately. Look from behind, they are as alike as two boats.

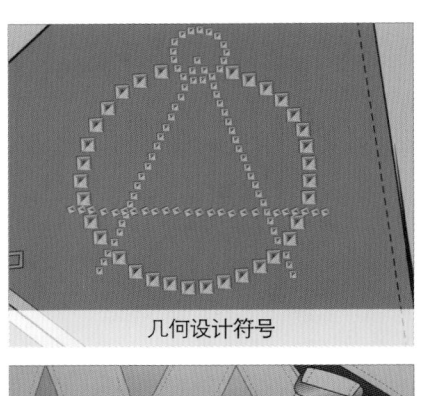
几何设计符号

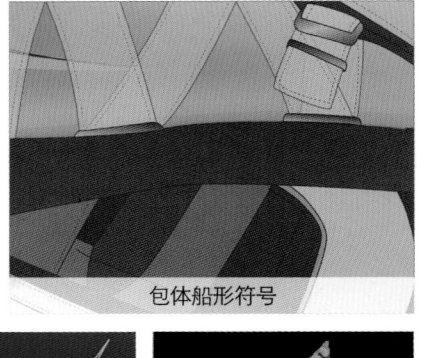
包体船形符号

包面湖面造型

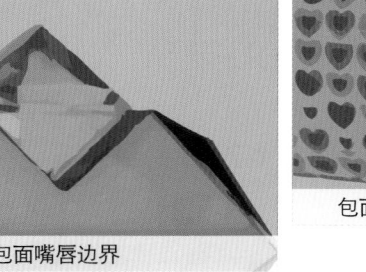
包面嘴唇边界

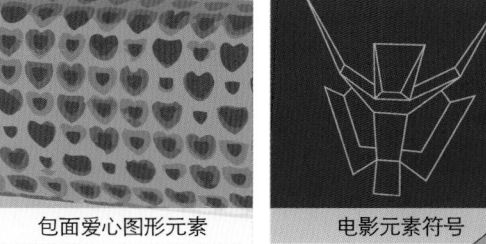
包面爱心图形元素

电影元素符号

星星符号

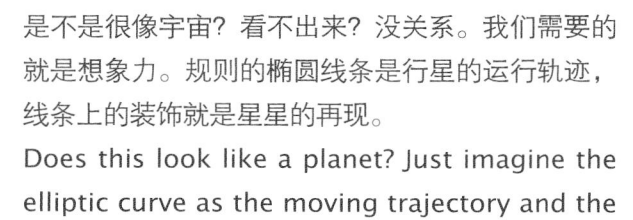

是不是很像宇宙？看不出来？没关系。我们需要的就是想象力。规则的椭圆线条是行星的运行轨迹，线条上的装饰就是星星的再现。

Does this look like a planet? Just imagine the elliptic curve as the moving trajectory and the deco along the curve is the planet.

北极冰川图形

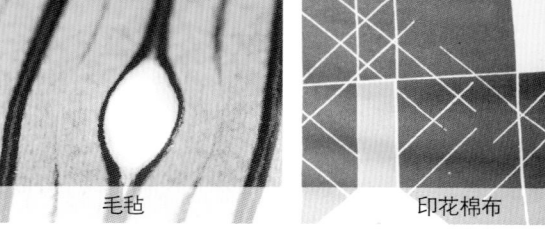

提花面料

印花PVC

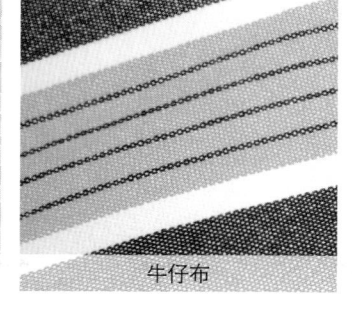
牛仔布

毛毡

印花棉布

S 箱包

2015 春夏海派时尚流行趋势

时空筑梦
REALISING DREAMS THROUGH SPACE

有机金属形态 　　　　节点设计 　　　　延伸框架 　　　　晶体结构

动物骨骼 　　　　鸟巢屋顶　　　　动态车身

设计灵感

仿盔甲设计既坚固又耐看，金属色透射出机械的冷酷感。
Inspired by armory, this metallic tone design is both functional and cool.

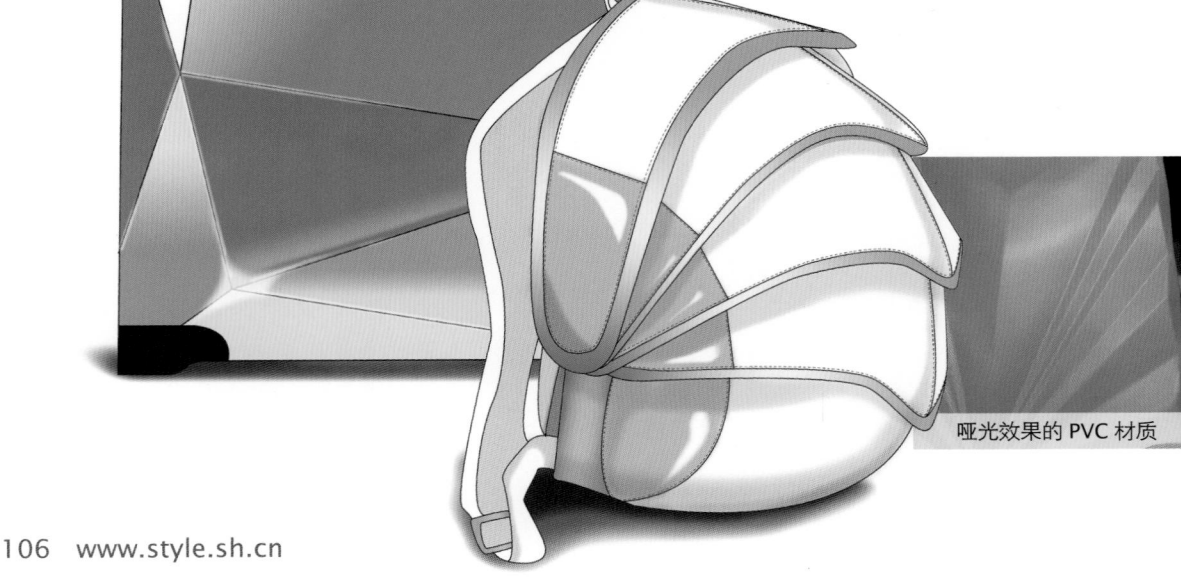

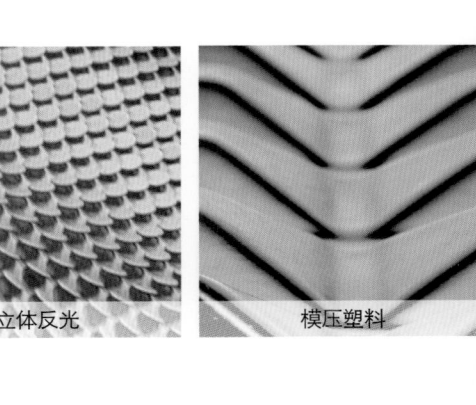

哑光效果的 PVC 材质　　　　立体反光　　　　模压塑料

106　www.style.sh.cn

2015 SPRING/SUMMER STYLE SHANGHAI FASHION TREND / BAG&SUITCASE

关 键 要 素 | 机械生物
KEY POINTS | MECHANICAL CREATURE

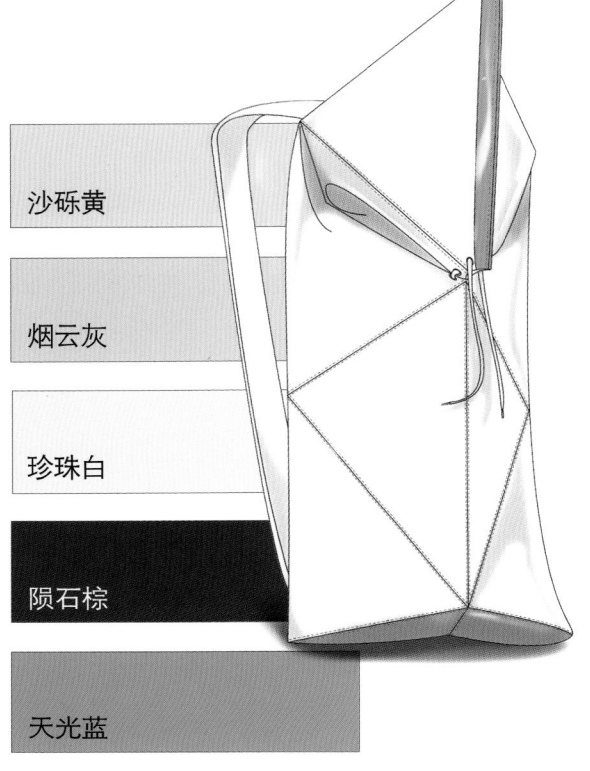

沙砾黄

烟云灰

珍珠白

陨石棕

天光蓝

立体烟囱式设计使包袋具有动态美学，块面分割使得包袋极具观赏性。
Patches of materials add dynamics to this chimney shaped bag.

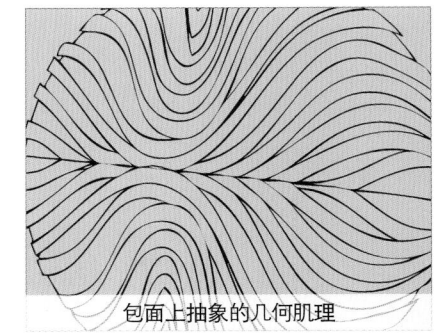

包面上抽象的几何肌理

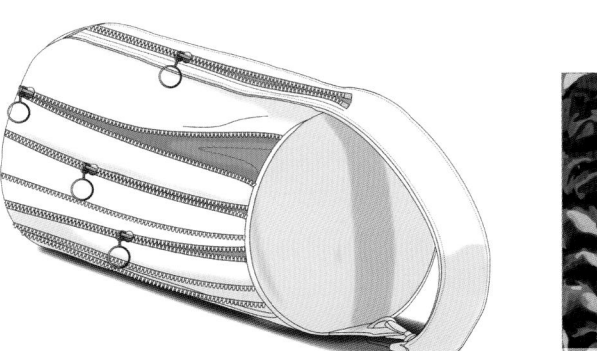

包面奇幻生物的仿生

体块感在包面上的表现

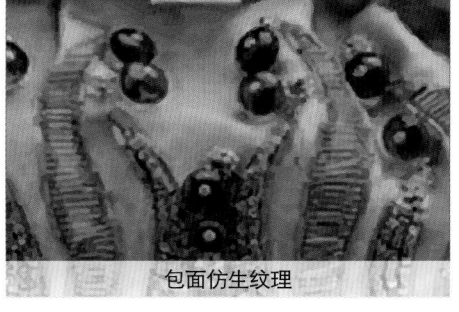

包面仿生纹理

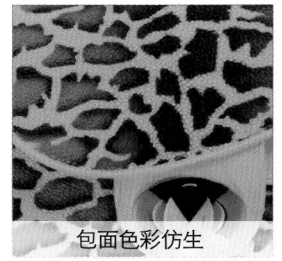

包面色彩仿生

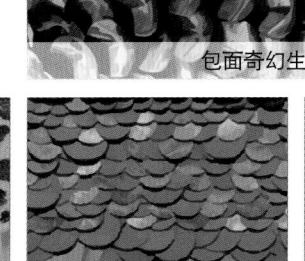

包面鳞片材质仿生

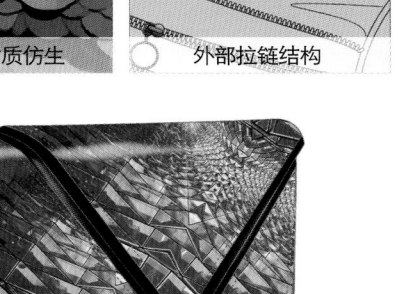

外部拉链结构

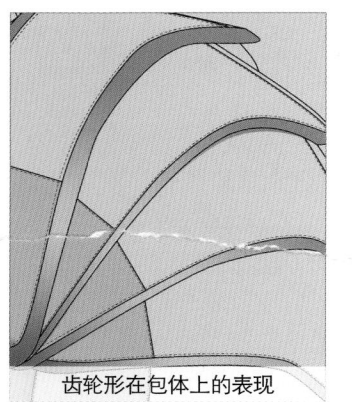

分割块面感和肌理面料的设计组合，形成块面碰撞的效果。
Blocks and contrasting metallic materials.

齿轮形在包体上的表现

金属扣件

印花皮革

水晶饰布

PU材料

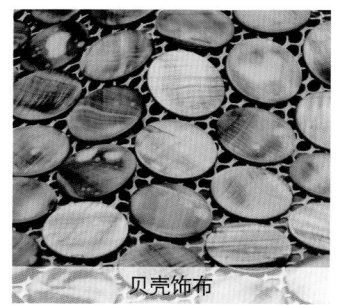

贝壳饰布

S箱包 2015 春夏海派时尚流行趋势

时空筑梦
REALISING DREAMS THROUGH SPACE

玻璃球

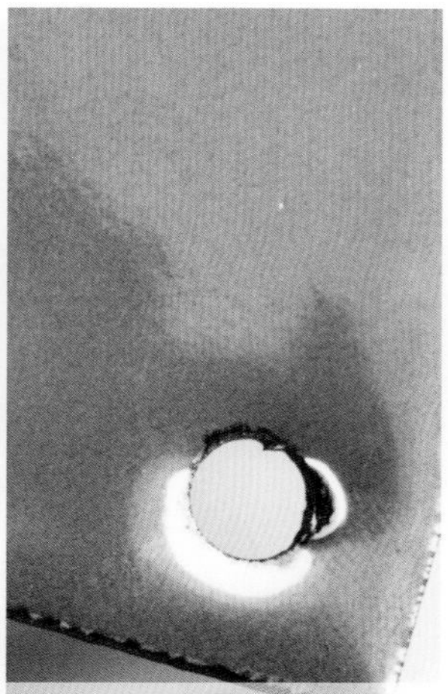

金属片

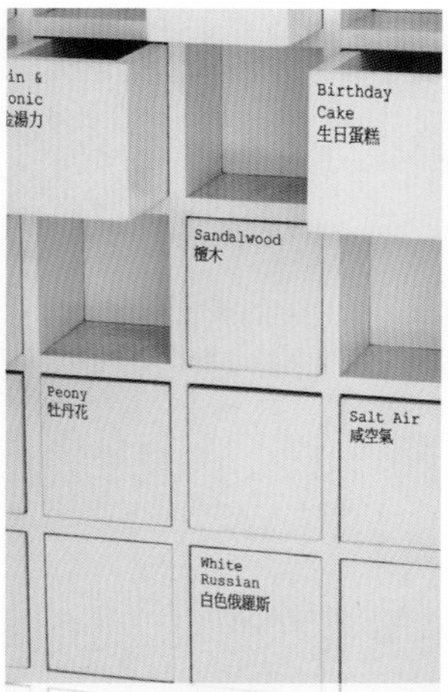

储物架

白灯

纯白的汽油桶

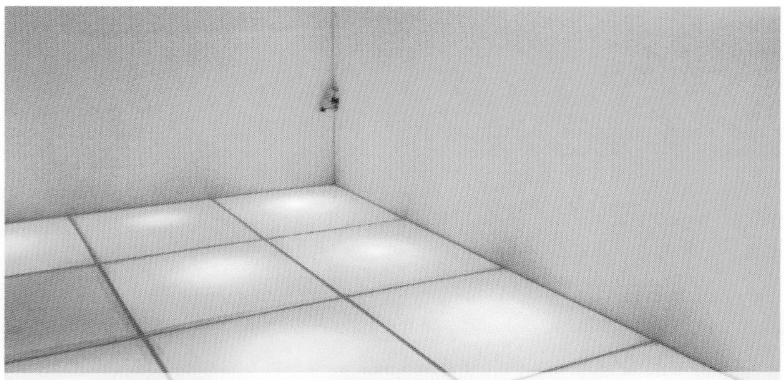

单纯的墙角

白墙壁

设计灵感

箱包设计以未来材质为主。在女式手提包上，以透明 PVC 为材质，营造空间错乱感，旅行箱上多用镜面纹理表现未来单纯感。
Materials are our main concern. We use PVC on women's handbags and shiny polycarbonate on luggages.

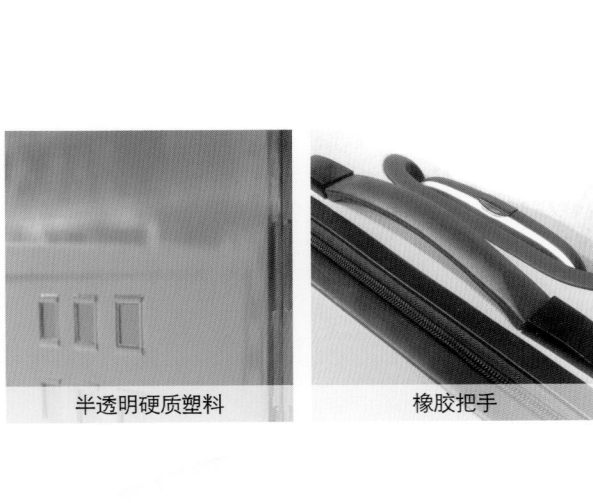

半透明硬质塑料

橡胶把手

柔韧薄木片

108　www.style.sh.cn

2015 SPRING/SUMMER STYLE SHANGHAI FASHION TREND / BAG&SUITCASE

关键要素 | 单纯设计
KEY POINTS | SIMPLE DESIGN

烟云灰

氯化灰

珍珠白

日光白

碳钢灰

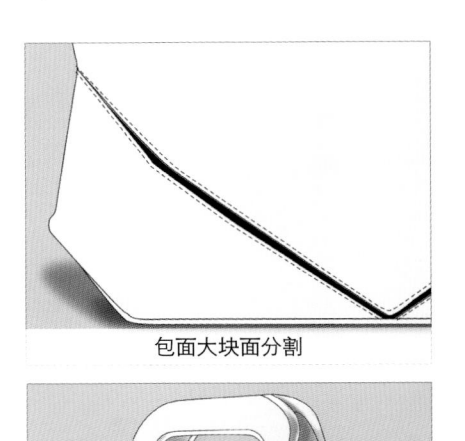

包面大块面分割

极简把手设计

女式包袋的单纯设计主要表现在线条分割上。"门"字形简单包面、黑色对比手柄，干净大气。
Inspired by Chinese character "门" (means "door"), we get this sleek silhouette and with a contrasting black handle, the design speaks itself.

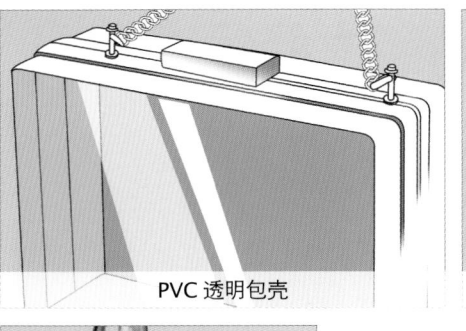

PVC 透明包壳

大体块包型

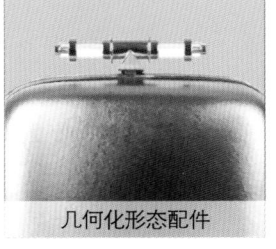

几何化形态配件

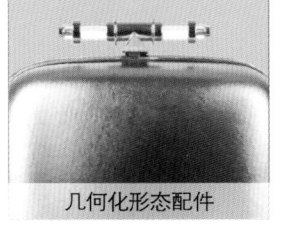

PV 金属亮光

简单的插口结构

包面漆皮反光

这款配件设计运用白色硬壳塑料，包纯色边，直接表现干净单纯的感觉。
New processing techniques allow us to make beautiful accessories that couldn't be made only with hands. This white hard-shell case with pure color trimming looks very neat.

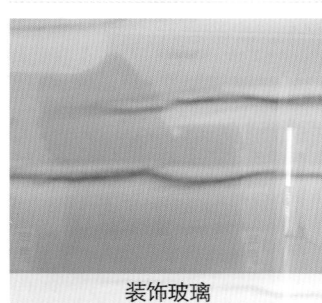

装饰玻璃

粘接皮革板

牛皮

金属铆钉

尼龙拉链

 **箱包** 2015 春夏海派时尚流行趋势

时空筑梦
REALISING DREAMS THROUGH SPACE

干净的空间

简单设计

透明线条

透明线条

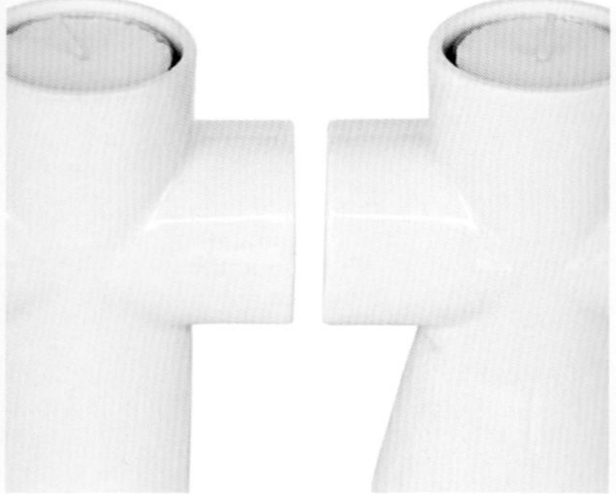

白色对称

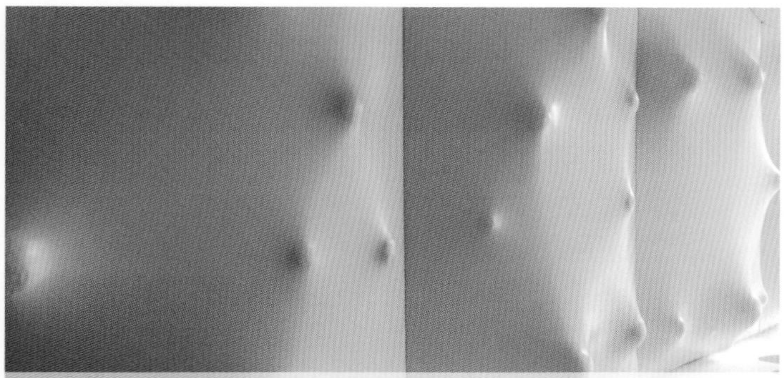

光线照射

简单块面

设计灵感

干净的设计无需过多的颜色，细微之处见设计。左边的包款采用大体块面设计，两边沿用细节绑带设计，简约大气。

Less is more. Large panels splice texture into a classic silhouette with on braided, chain-link straps on both sides.

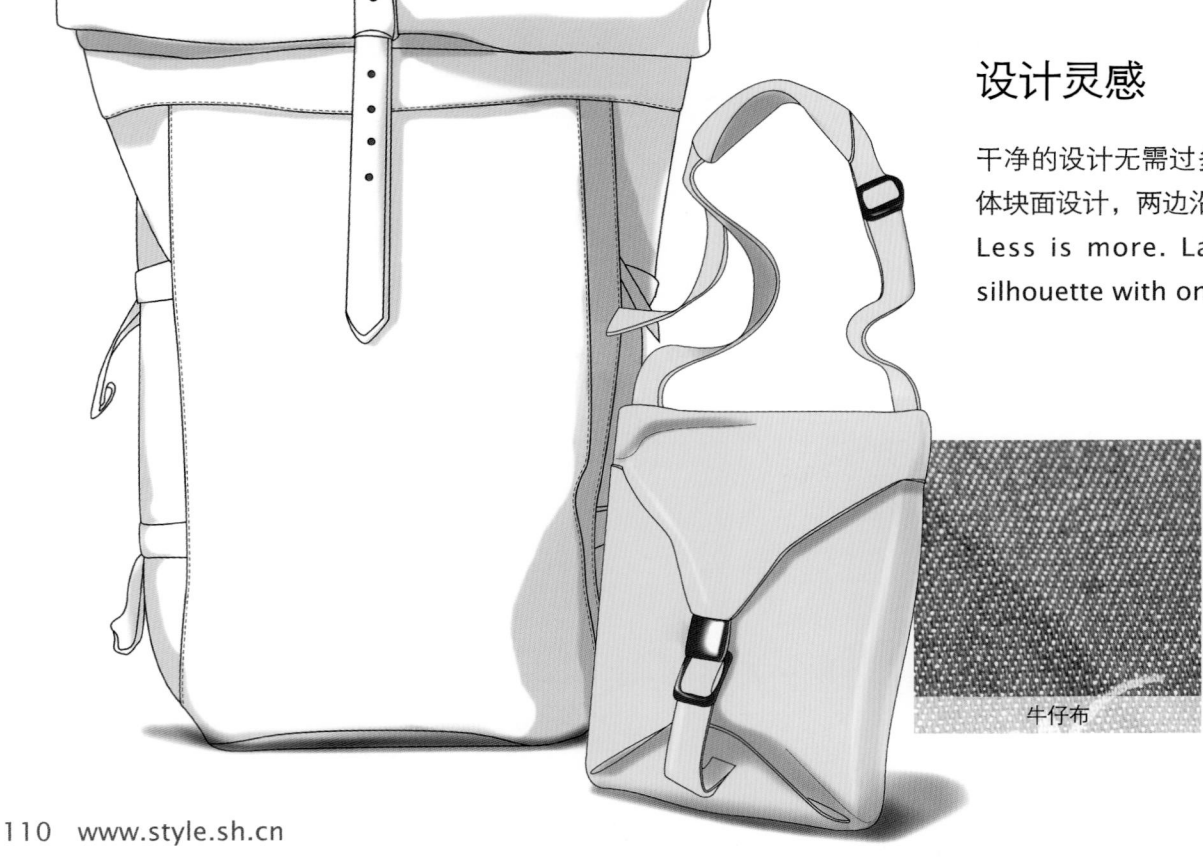

牛仔布

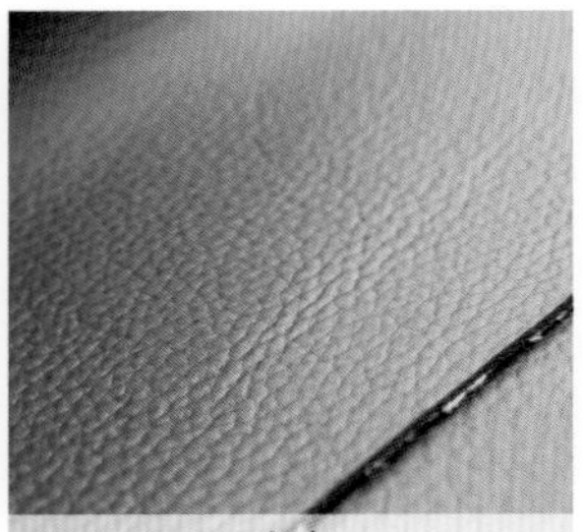

牛皮

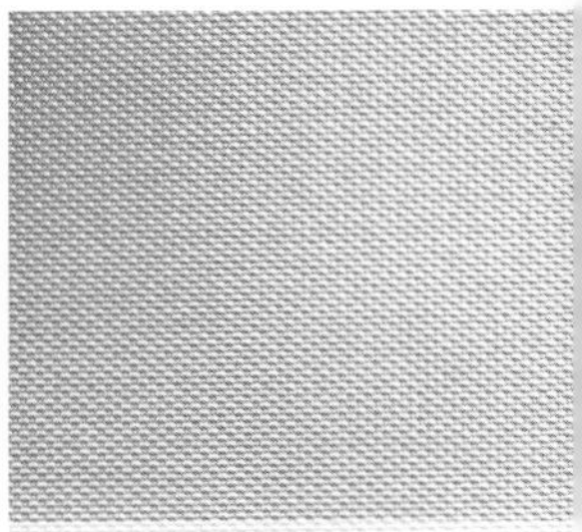

牛津布

2015 SPRING/SUMMER STYLE SHANGHAI FASHION TREND / BAG&SUITCASE

关 键 要 素 | 单 纯 设 计
KEY POINTS | SIMPLE DESIGN

烟云灰

氯化灰

珍珠白

日光白

碳钢灰

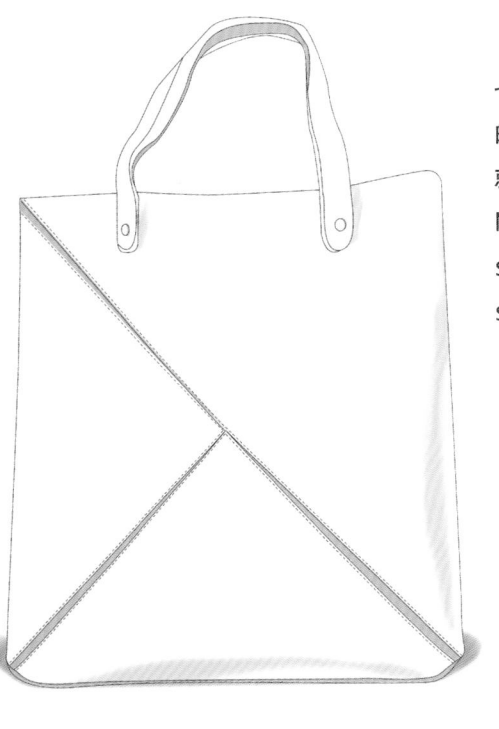

七巧板的舍弃型设计。去除多余的细节，还原皮质的块面体现，就是单纯设计。
No unnecessary design. Panels splice different texture into a sleek silhouette.

包面单色亮色的漆皮材质

透明 PVC 材质的包体

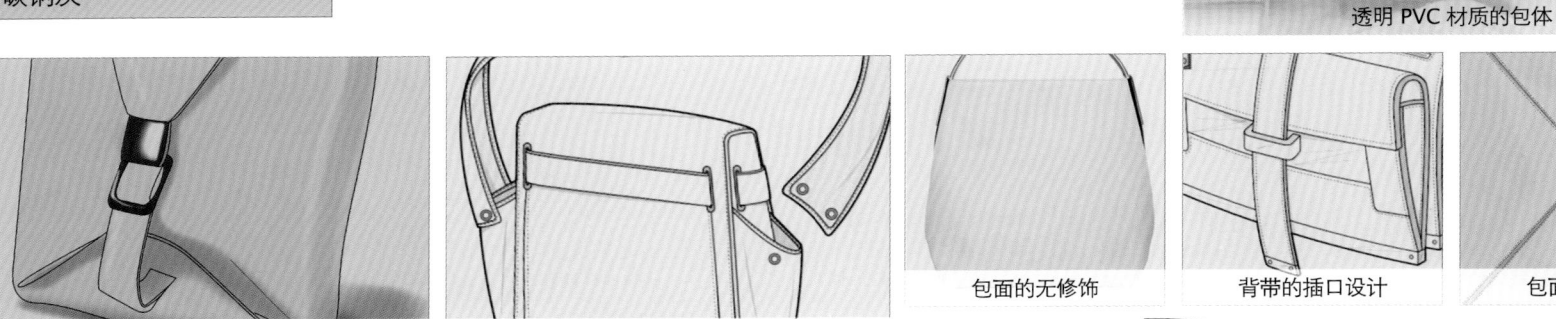

简单的包体结构

背带的单纯设计

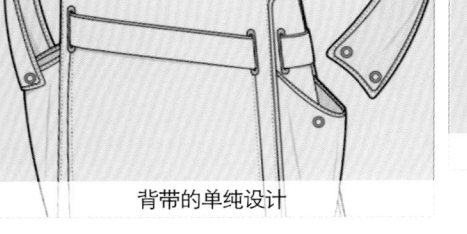
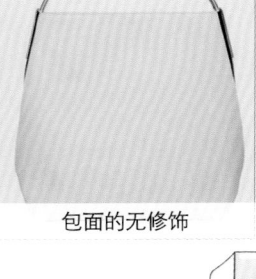

包面的无修饰

背带的插口设计

包面折叠设计

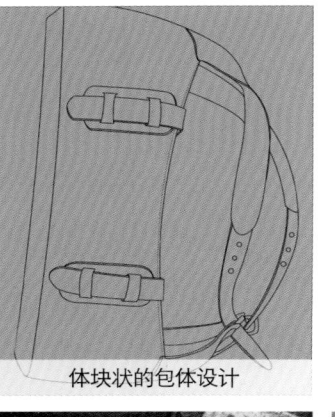

体块状的包体设计

斜挎包的设计简单干净，搭扣采用一字型襻扣，既可斜背亦可正背。
Beautifully finished edges, shining nickel hardware and an optional cross body strap perfect the chic, elegant style.

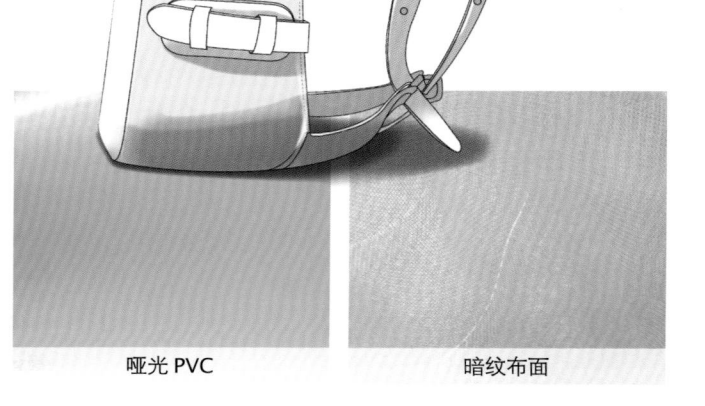

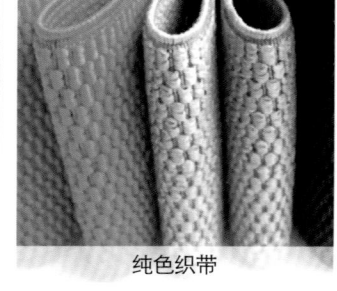

D 型扣

纯色织带

斜纹布

哑光 PVC

暗纹布面

 箱包 2015 春夏海派时尚流行趋势

多彩享梦
ENJOY COLORFUL DREAMS

老上海的生活呈现一片古典而又繁华的豪世盛景，将老上海怀旧经典而又不失小资生活的高贵气质元素融入在各类包的设计中，抒发人们对那个年代高雅情怀的怀念。

Life in old Shanghai was classical and bustling. We add some old temperament and nostalgia into various bags showing our respect to that era.

BAG&SUITCASE

2015 SPRING/SUMMER STYLE SHANGHAI FASHION TREND
海 派 箱 包 流 行 趋 势

2015 SPRING/SUMMER STYLE SHANGHAI FASHION TREND / **BAG&SUITCASE**

关键要素 | KEY POINTS

童年拾趣
CHILDHOOD REFLECTIONS

跨界实验
TRANSBOUNDARY EXPERIMENTS

快乐崇拜
JOLLY IDOLATRY

都市街头
CITY STREETS

浅裸粉

蜜桃粉

幻彩红

冰山蓝

玛瑙蓝

松石绿

珊瑚红

迎春黄

琵琶橙

石榴红

S箱包 2015 春夏海派时尚流行趋势

多彩享梦
ENJOY COLORFUL DREAMS

拼图

彩色糖果

彩色水管装置设计

纺织面料设计中的几何图形

儿童画

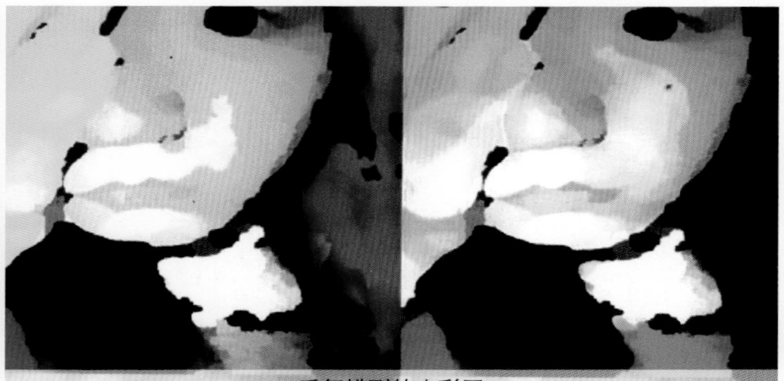

重复排列的水彩画

装饰灯具

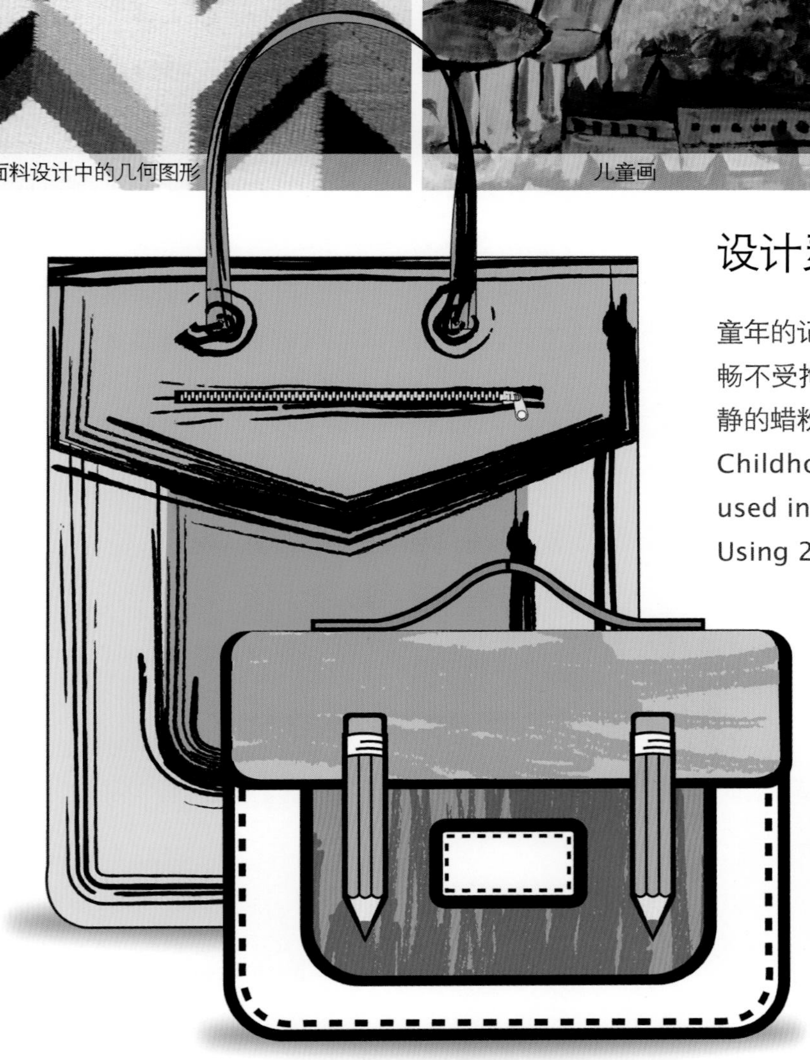

设计灵感

童年的记忆是美好的，将儿童画里动态的画笔效果用在简单的托特女包设计中，显得流畅不受拘束，快乐自由。受到 2D 和 3D 视觉转换的启发，女包设计融入了动感而又沉静的蜡粉色彩，为这一季的女包设计带来新的活力。

Childhood brings us too much happy memory. Inspired by brushes usually used in children's cartoon book, this handbag has a dynamic and free look.
Using 2D graphic design for 3D illusion creates a fresh look on this bag.

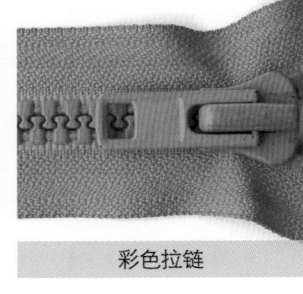

彩色拉链

牛津布

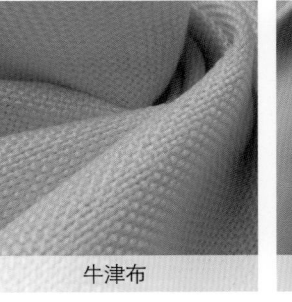

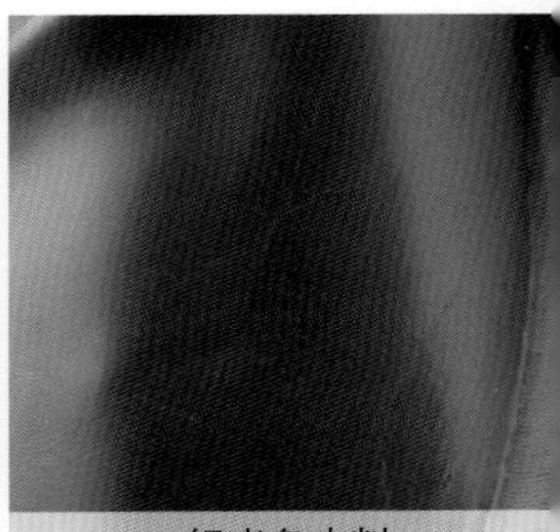

银光色皮料

2015 SPRING/SUMMER STYLE SHANGHAI FASHION TREND / **BAG&SUITCASE**

关 键 要 素 ｜ 童年拾趣
KEY POINTS ｜ CHILDHOOD REFLECTIONS

邮差包简单的造型深受年轻人的喜爱，时尚运动的感觉让这种款型成为了运动男的首选，其容量大的特点也为他们添色不少。
Featuring all of the functionality that an active life demands, a roomy messenger bag features durable construction and effortless organization.

泳池蓝

迎春黄

珊瑚红

幻彩红

霓虹绿

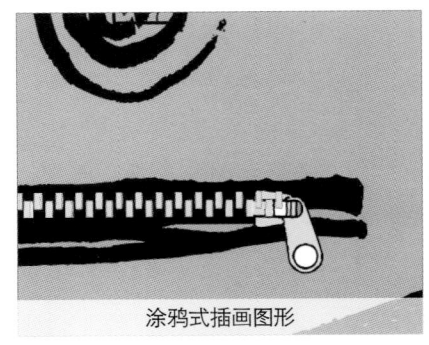

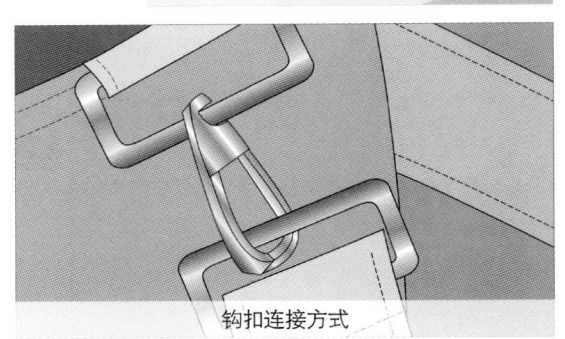

涂鸦式插画图形

钩扣连接方式

粗犷的假缝线装饰

侧边安装肩带

肩带上的插扣结构

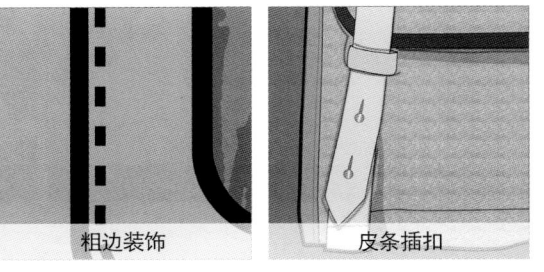

粗边装饰

皮条插扣

卡通图形装饰

将童年游戏飞行棋中的箭头进行排列，作为包袋上的装饰，显得时尚新潮，感觉回到了童年。
Arrays of colorful arrows in flight chess make this bag stylish and fashionable reminding us happy childhood.

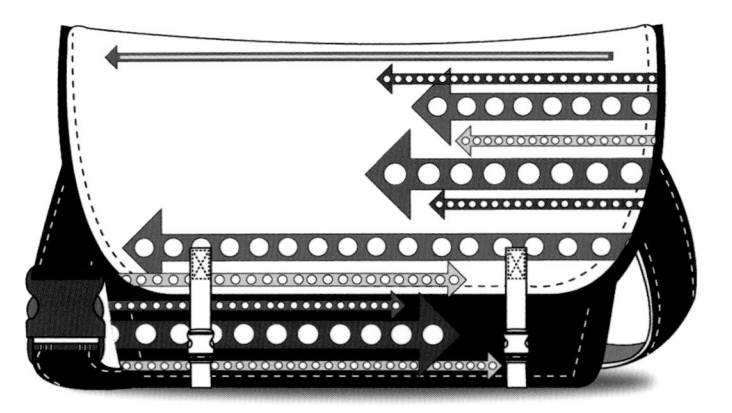

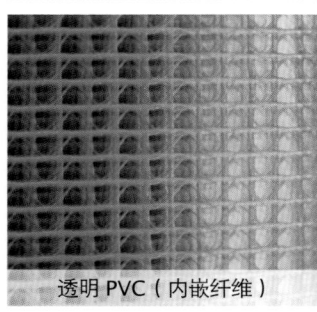

透明 PVC（内嵌纤维）

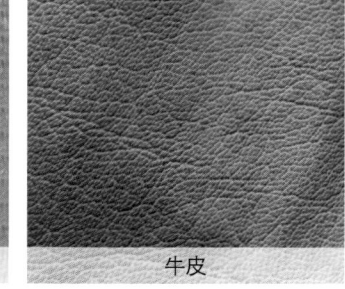

牛皮

塑料插扣

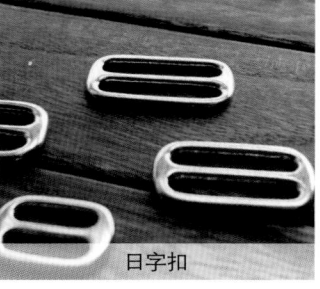

日字扣

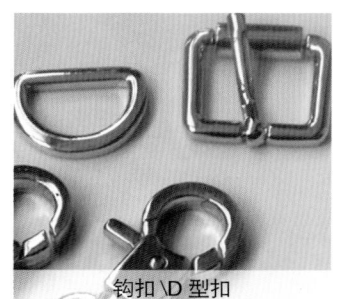

钩扣\D 型扣

115

S 箱包 2015 春夏海派时尚流行趋势

多彩享梦
ENJOY COLORFUL DREAMS

彩色药片组合成的装饰画

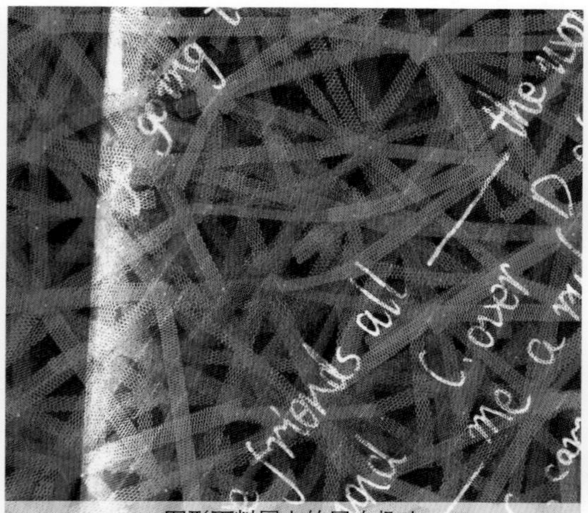

图形面料展上的层次趣味

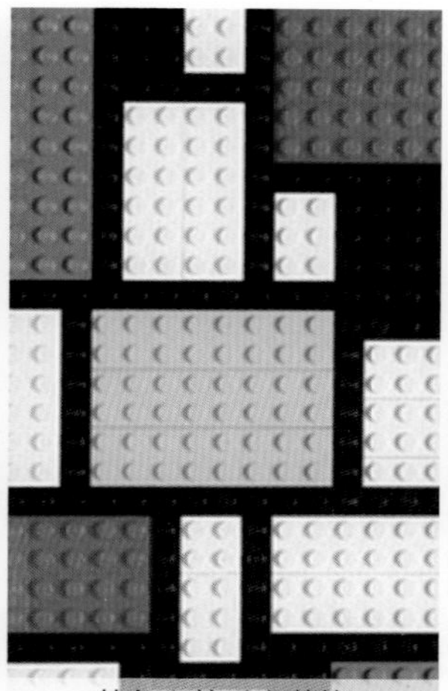

箱包上的乐高装饰

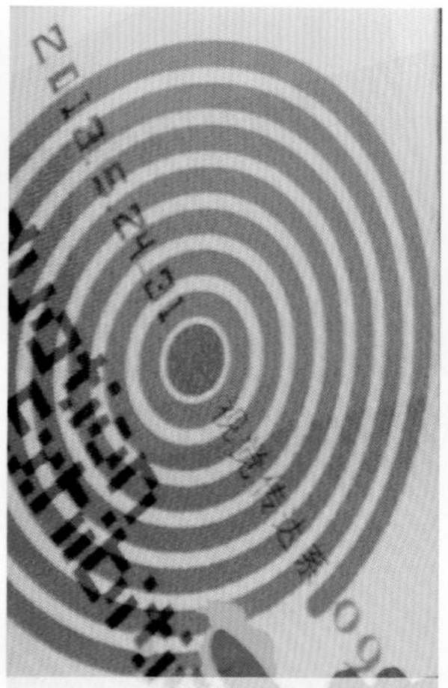

海报设计

平面设计中的节奏趣味

乐高积木

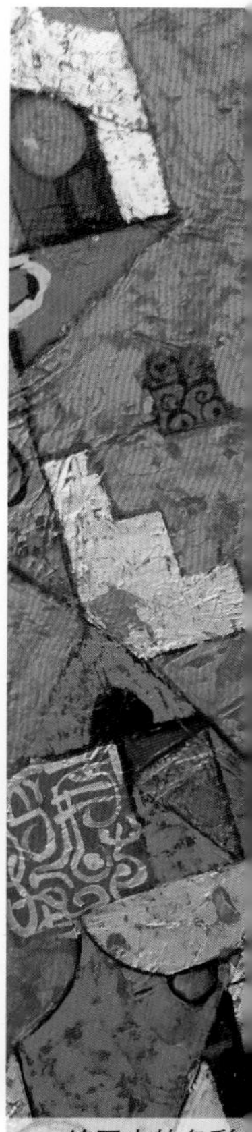

绘画中的色彩

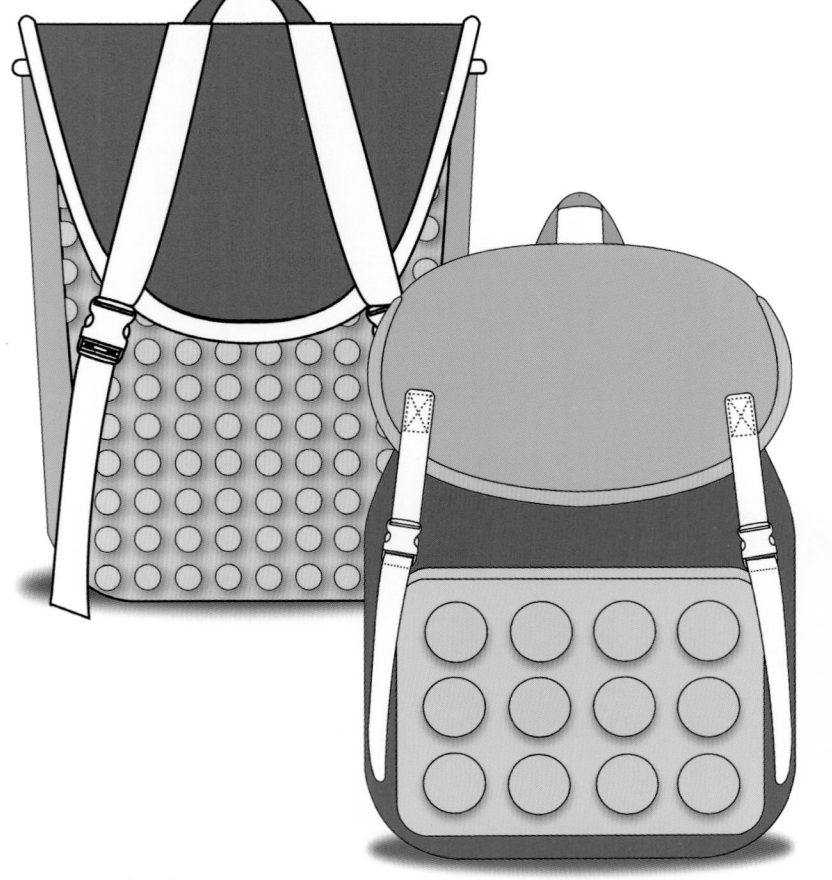

设计灵感

乐高玩具充满了各种奇妙的组合，男女老少都能在其中寻找到美好的记忆。在儿童书包设计中运用一些乐高的元素，给生活增添无穷的乐趣。Lego encourages people of different age to construct things with colorful bricks and plates. Inspired by Lego, we also encourage kids to have fun in study.

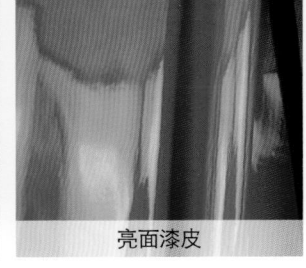

彩色尼龙布

亮面漆皮

白色尼龙布

116　www.style.sh.cn

2015 SPRING/SUMMER STYLE SHANGHAI FASHION TREND / **BAG&SUITCASE**

关键要素 | 童年拾趣
KEY POINTS | CHILDHOOD REFLECTIONS

松石绿

石榴红

霓虹绿

迎春黄

泳池蓝

柔软的皮质运动包呈现出露营包形状，用一些缎带、图案色彩或图形彰显童年天真趣味的生活状态。

Some faille straps and colorful patterns mark the front of a sizable gym bag crafted in supple leather for a handsome look.

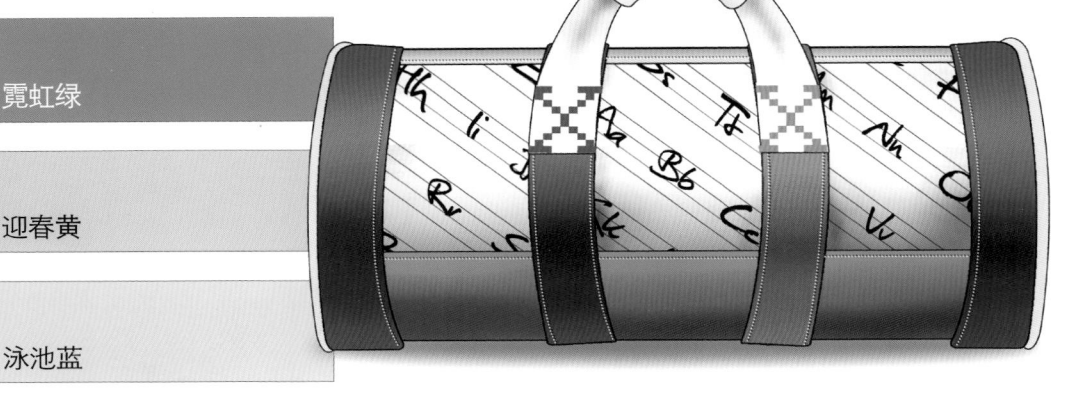

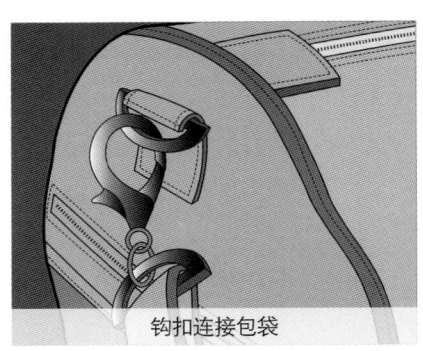
钩扣连接包袋

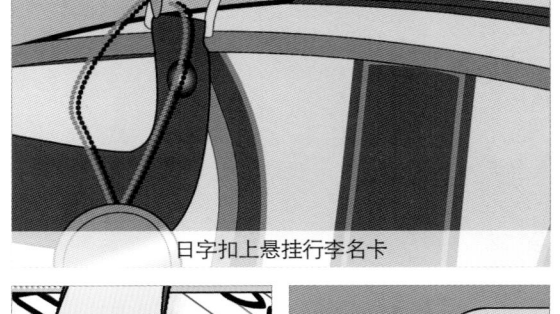
日字扣上悬挂行李名卡

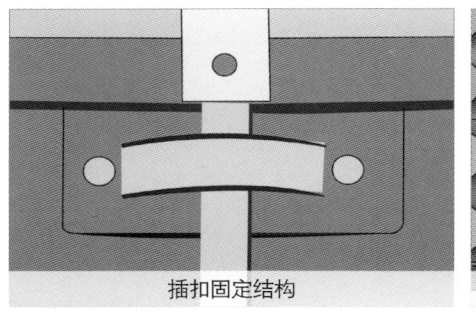
插扣固定结构

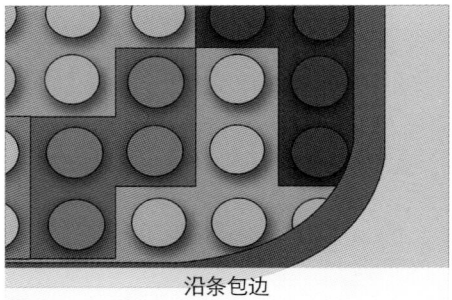
沿条包边

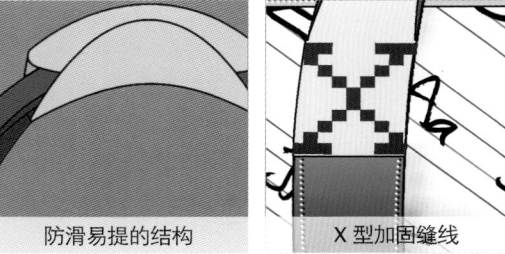
防滑易提的结构

X 型加固缝线

箱包侧袋

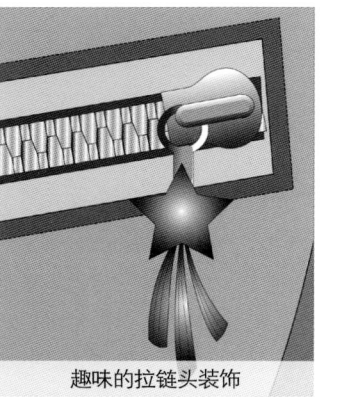
趣味的拉链头装饰

鲜艳的皮革和花纹面料结合，既强化了视觉形象，又勾起了对过往的记忆。

Bright tone leather and interesting prints define this eye-catching sports bag.

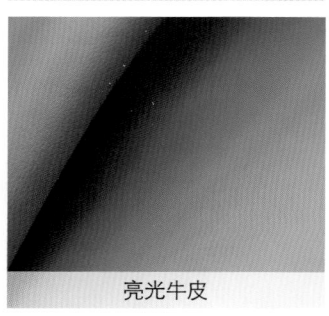
亮光牛皮

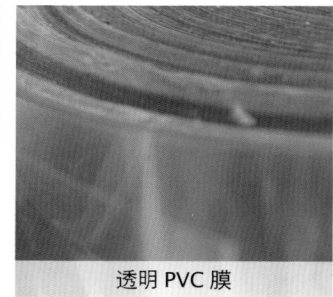
透明 PVC 膜

金属拉链

软硅胶乐高锁扣

粗料插扣

117

S 箱包

2015 春夏海派时尚流行趋势

多彩享梦
ENJOY COLORFUL DREAMS

乐高积木展上的乐高恐龙

橱窗展示道具 | 橱窗瓷器

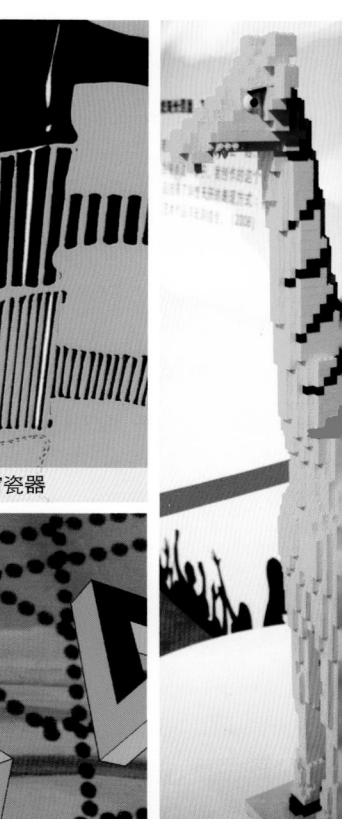
乐高玩具展

恐龙骨骼图形设计

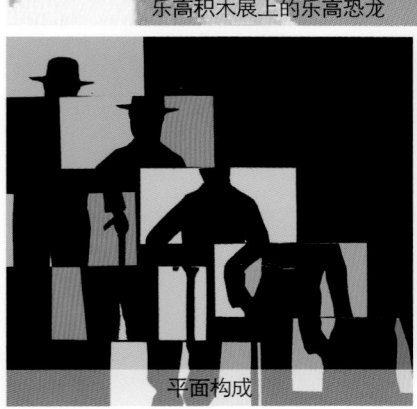
平面构成

图形设计中的矛盾空间

设计灵感

将传统印象中的动物轮廓，利用视错觉的表达形式尝试新颖的图像式印花，同时增添视觉趣味，让跨界设计带来新的情感表达。

Inspired by abstract animal shapes, these patterns intend to catch your attention and bring different feelings.

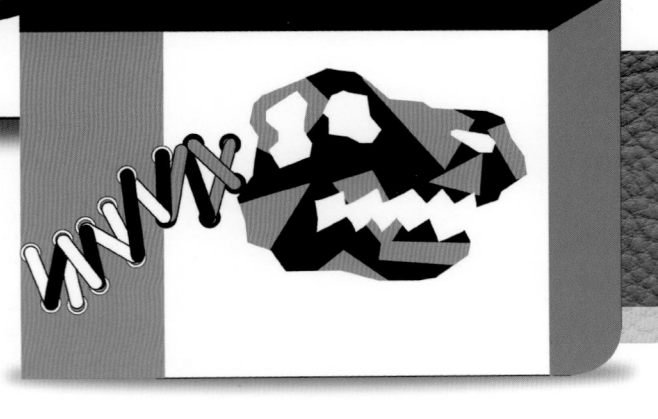

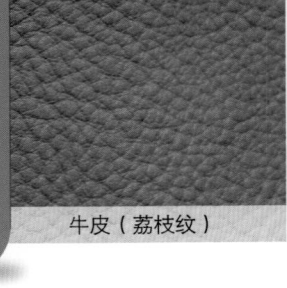
牛皮（荔枝纹）

蛇皮

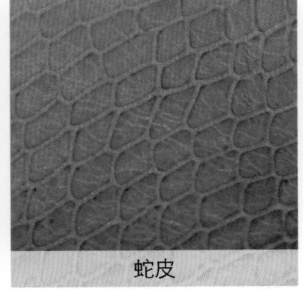
光面牛皮

118   www.style.sh.cn

2015 SPRING/SUMMER STYLE SHANGHAI FASHION TREND / BAG&SUITCASE

关键要素 | 跨界实验
KEY POINTS | TRANSBOUNDARY EXPERIMENTS

泳池蓝

幻彩红

迎春黄

松石绿

乌梅黑

通过包体外印刷另一个包型，造成视觉上以为包的里面还有另外一个包。而彩色的立体图形和黑色的对比又让包变得更加多彩、靓丽。
Inspired by "Together bag", we use more colorful 3D patterns and contrasting black to furnish it.

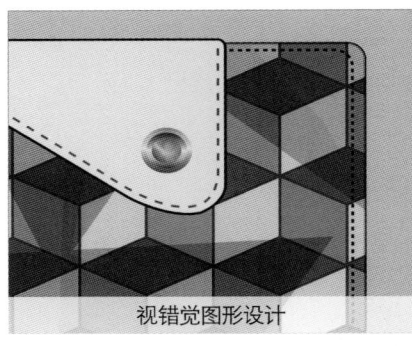

视错觉图形设计

穿插绑带骨骼结构

内部设计结构

顶部开口拉链

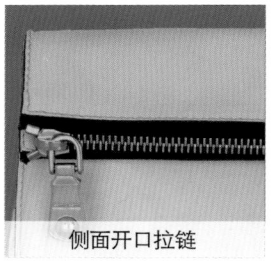

侧面开口拉链

图形化动物形态

三维立体空间图案

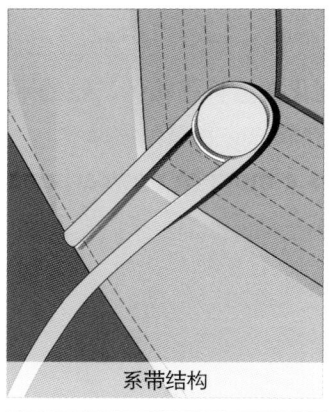

系带结构

将具有矛盾空间、视错觉图形的面料作为主要面料，让图形产生与客观事实不相符的空间效果。
Use graphic design to create illusion on the bag.

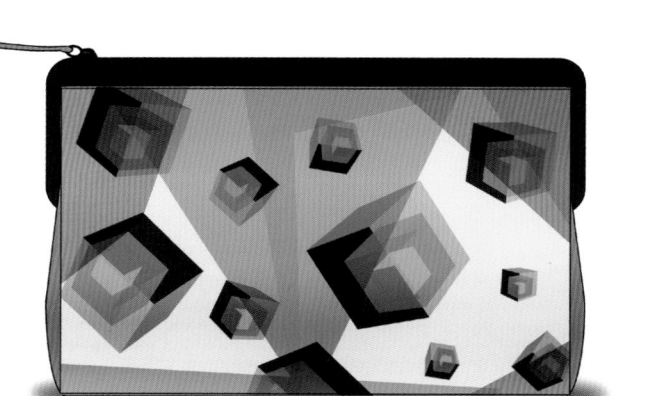

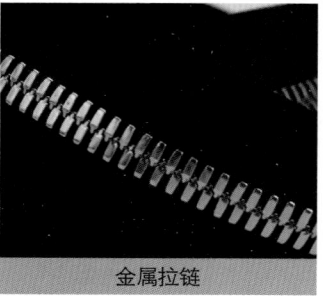

金属拉链

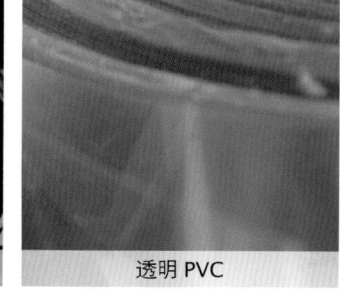

透明 PVC

气眼

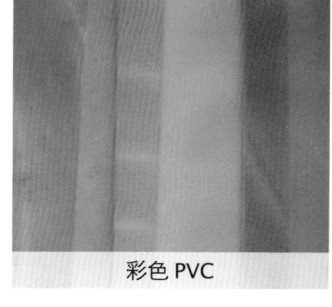

彩色 PVC

有色织带

119

箱包 2015 春夏海派时尚流行趋势

多彩享梦
ENJOY COLORFUL DREAMS

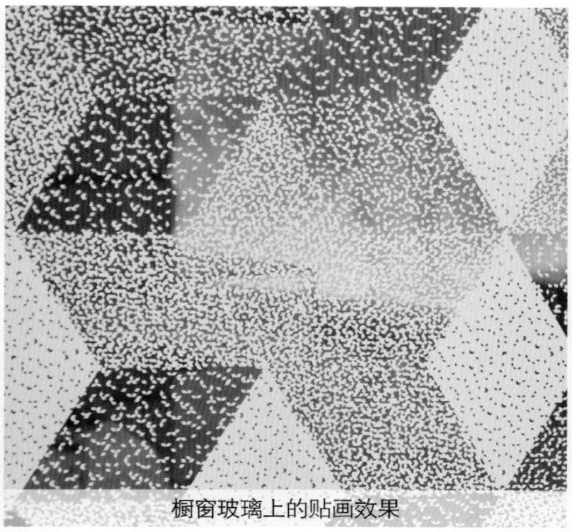

橱窗玻璃上的贴画效果

七巧板拼色

立体构成练习

玻璃的镜面装饰

灯具设计中的镂空图形

座椅上的几何图形

立体空间图形

设计灵感

利用视点的转换、交替和图形色彩的变化在拉杆箱箱体二维的平面上表现出三维的立体形态，从而达到跨界表达的效果。

Graphic design of 3D illusion details a modern polycarbonate suitcase with different visual effects.

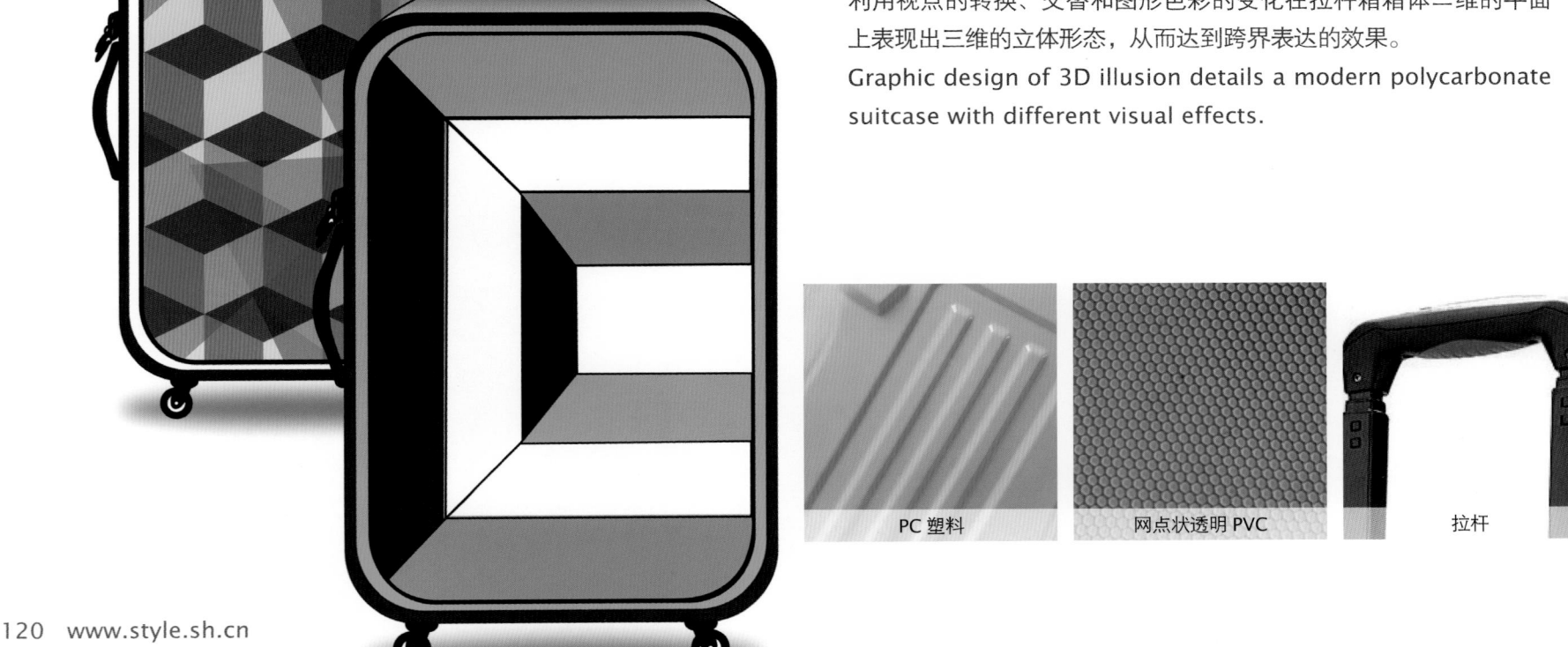

PC 塑料

网点状透明 PVC

拉杆

120　www.style.sh.cn

2015 SPRING/SUMMER STYLE SHANGHAI FASHION TREND / BAG&SUITCASE

关键要素 | 跨界实验
KEY POINTS | TRANSBOUNDARY EXPERIMENTS

珊瑚红

幻彩红

浅草绿

泳池蓝

松石绿

将平面矛盾空间的表达方式结合到箱包面料的设计中，增强了箱包的艺术性，引起了对箱包的另外一种趣味性的理解。
Inspired by contradiction space, we hope these handbags and luggage have artistic quality.

图案的拼接

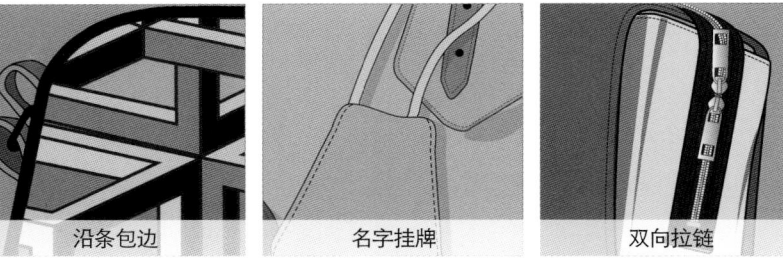

精致把手

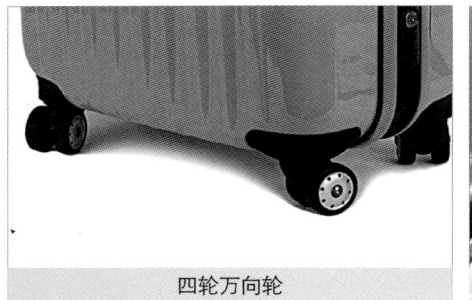

四轮万向轮

铝框结构

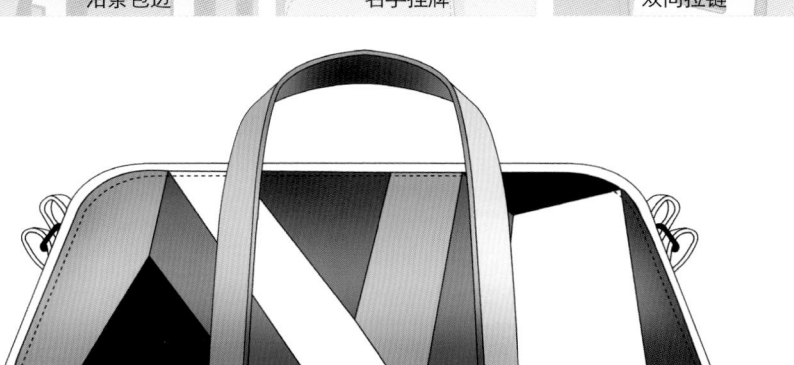

沿条包边

名字挂牌

双向拉链

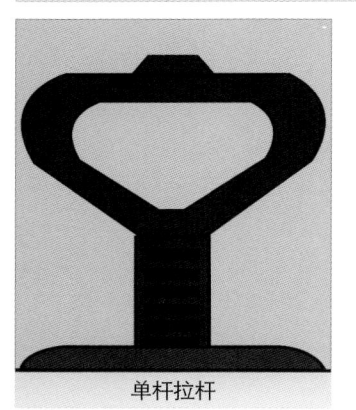

单杆拉杆

T型运动包持续以往简约大气的廓型，通过色彩、空间变化打造与众不同的视觉冲击力。
Various colors and shapes add vitality to this T-shaped gym bag.

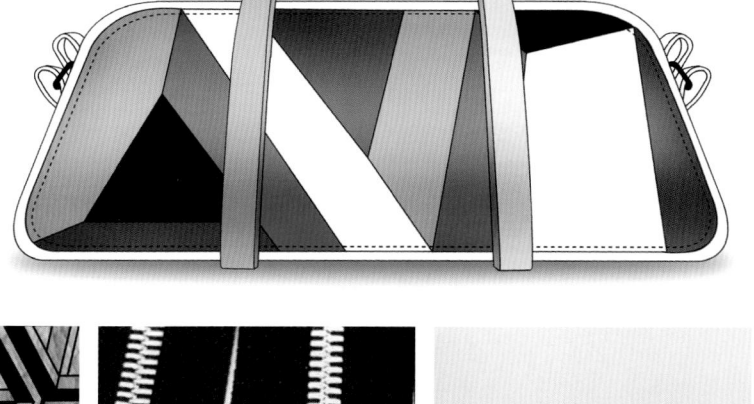

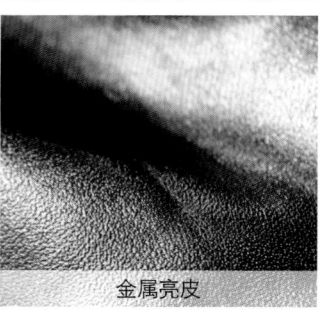

金属亮皮

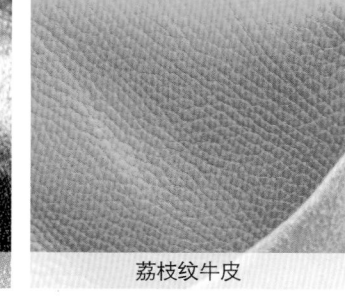

荔枝纹牛皮

印有立体图形的牛皮

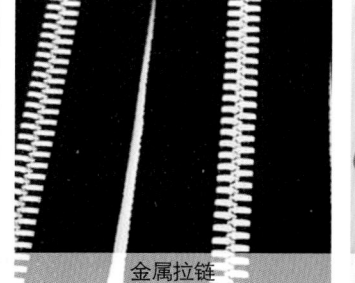

金属拉链

蘑菇钉

S箱包 2015 春夏海派时尚流行趋势

多彩享梦
ENJOY COLORFUL DREAMS

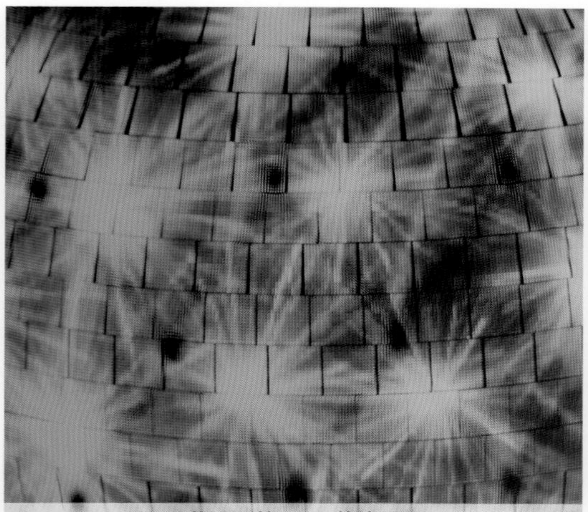

世博园的 LED 装饰球

某酒店大堂的装饰灯

电影博物馆外墙夜景

电影博物馆灯光场景

球体上变化的图像

地面灯光投影

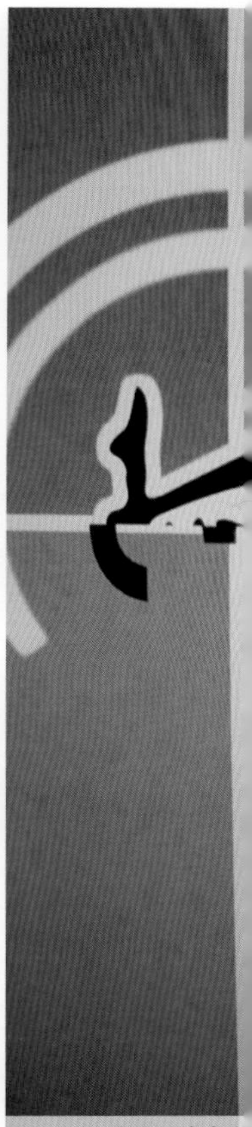

电子倒计时

设计灵感

在女包设计上，运用透明的 PVC、简约的造型打造小提包圆润的结构感，信封包袋采用平滑流畅的柔软皮革，适合日常使用。条带可拆卸或可调节。甜美的粉蜡色彰显优雅，符号型图形则为手提斜跨包打造生动、现代的感觉。

Curvaceous detailing ornaments the PVC crossbody bag. Sweet pastel shines on a soft leather messenger bag that exudes street-chic style. Removable and adjustable shoulder strap details further the polished design.

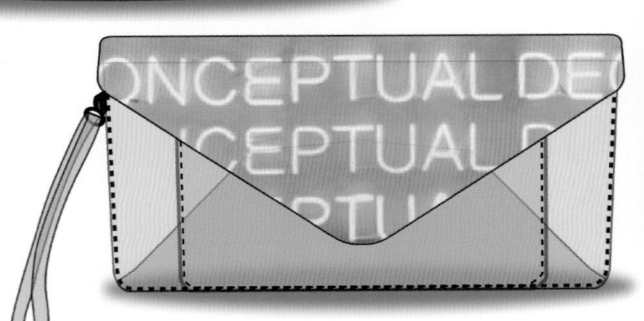

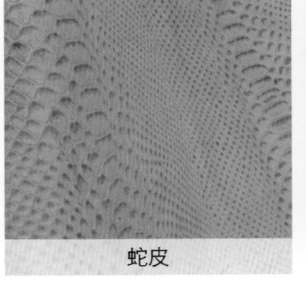

蛇皮

箱包钩扣

箱包 D 型扣

122 www.style.sh.cn

2015 SPRING/SUMMER STYLE SHANGHAI FASHION TREND / BAG&SUITCASE

关键要素 | 快乐崇拜
KEY POINTS | JOLLY IDOLATRY

幻彩红

深桃红

蜜桃粉

浅草绿

冰山蓝

半透明或着色塑料焕发现代感。包里的隔间袋无论是内置还是完全独立，都让包内的东西看来稍有庇护。
Transparent or frost plastic frame handbags either with an inner extra small bag or not expose your belongings with little coverage.

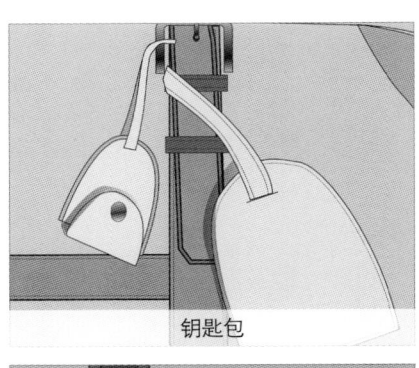

钥匙包

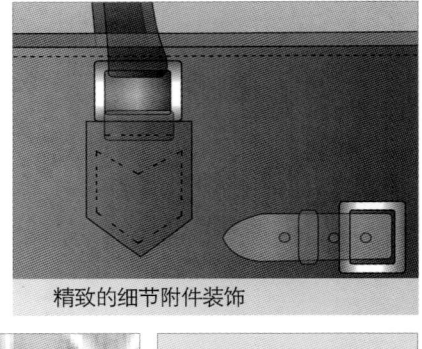

精致的细节附件装饰

行李名卡

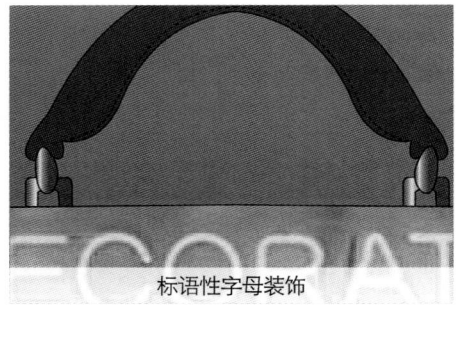

标语性字母装饰

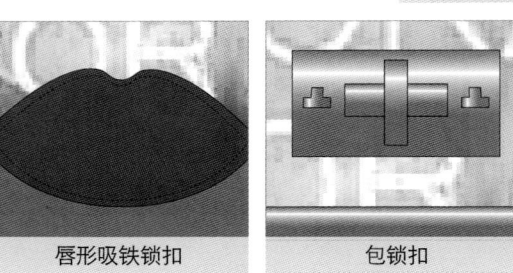

唇形吸铁锁扣

包锁扣

平钉的顺序排列

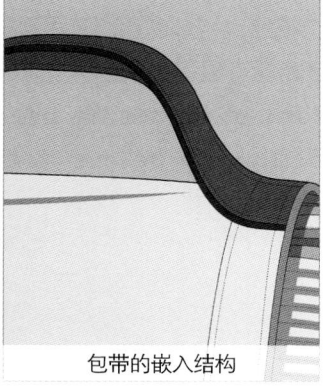

包带的嵌入结构

铆钉、半透明的材质、深邃的色彩、不对称的分割结构让包充满了神秘的色彩。
A sophisticated clutch with polished rivets will glam up your mysterious style.

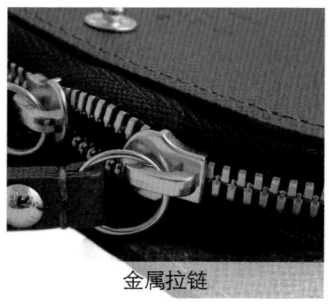

金属拉链

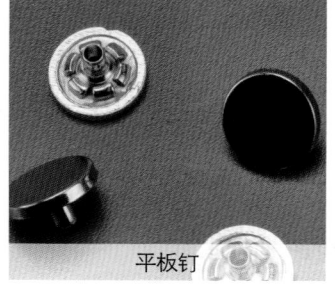

平板钉

毛毡（细腻）

光面牛皮

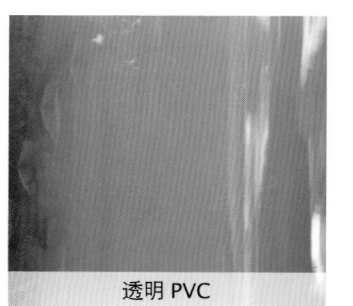

透明 PVC

箱包 2015 春夏海派时尚流行趋势

多彩享梦
ENJOY COLORFUL DREAMS

卡通漫画片段

纺织面料彩色几何图形设计

卡通彩色图案设计

木质墙面上字母涂鸦艺术

M50 创意园区 墙面涂鸦艺术（字母和动物图形组合）

都市贴标装

设计灵感

漫画创造了无数个英雄人物，这种都市英雄的情结也可以扩展地运用到箱包设计中，该类面料用在箱包的主体或者局部都是非常有吸引力的。

Comics such as Marvel have created tons of heroes. We use these characters or related elements in our design. We believe that these backpacks are very attractive and fashionable.

斑马纹面料

印花 PU 革

印花牛皮

124　www.style.sh.cn

2015 SPRING/SUMMER STYLE SHANGHAI FASHION TREND / BAG&SUITCASE

关键要素 | 都市街头
KEY POINTS | CITY STREETS

乌梅黑

石榴红

迎春黄

珊瑚红

松石绿

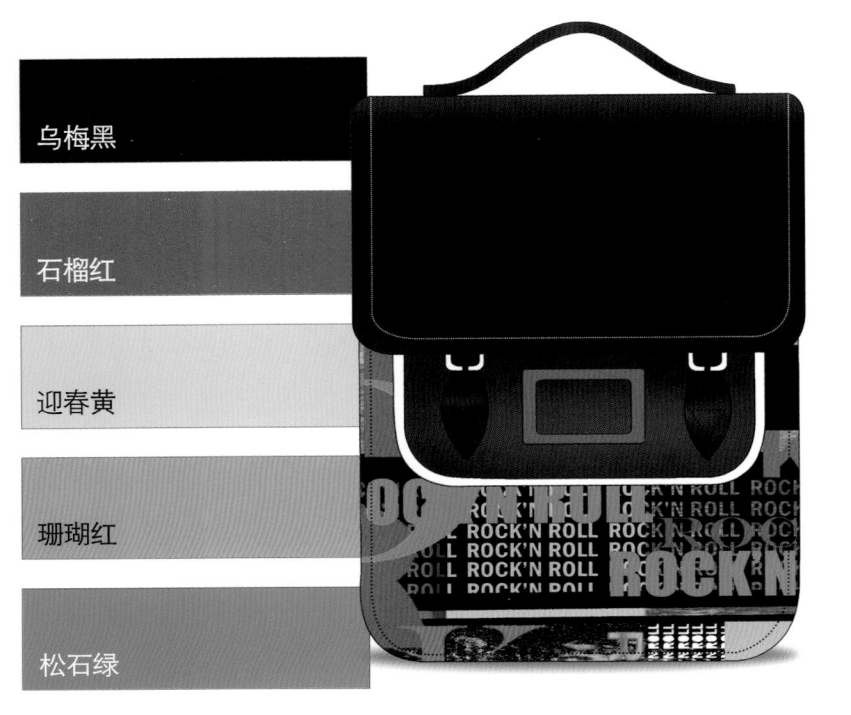

拉长学院包的比例，并搭配长长的双肩背带，彩色都市化的标语花布，让学院包又多了很多俏皮的感觉。
We widen original Cambridge flap satchel and add two back straps. Colorful cloth enlivens the design.

铆钉装饰和缝线吻合

假日字扣开合结构（吸铁结构）

三角形装饰固定侧袋

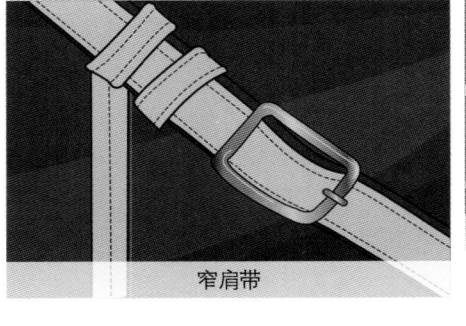

窄肩带

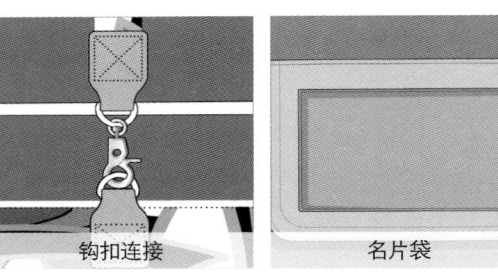

钩扣连接 　名片袋 　把手细节装饰

厚度松紧调整带

中等尺寸女学生提包用途广泛，适合各种场合，配上红色翻盖和点缀着可爱黑白斑马纹的包面令人眼前一亮。
A mid-sized satchel keeps you organized while on the go. Bright red flap and Exotic zebra stripes add marvelous mod style.

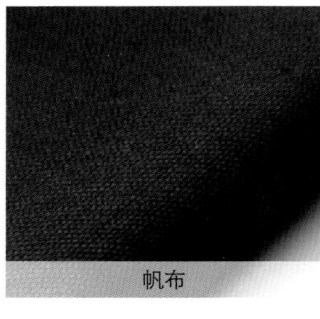

帆布

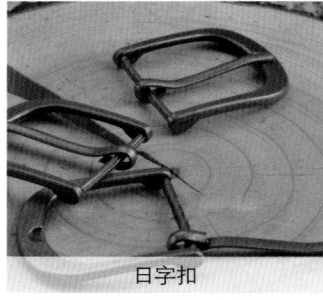

日字扣

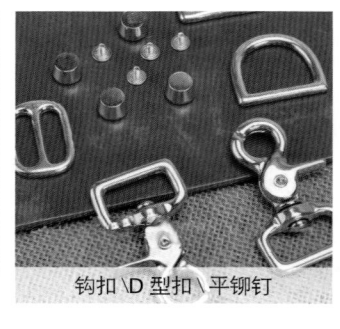

钩扣\D型扣 平铆钉

黑色PU革

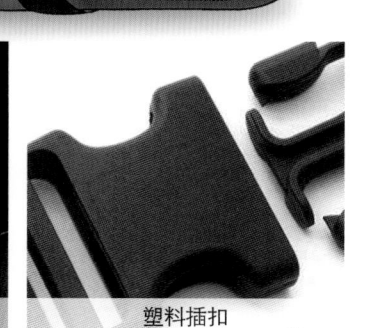

塑料插扣

125

S箱包 2015 春夏海派时尚流行趋势

多彩享梦
ENJOY COLORFUL DREAMS

乐高玩具的色彩层次及块面感

涂鸦墙面

彩色编织

折纸构成中的色彩以及体积变化

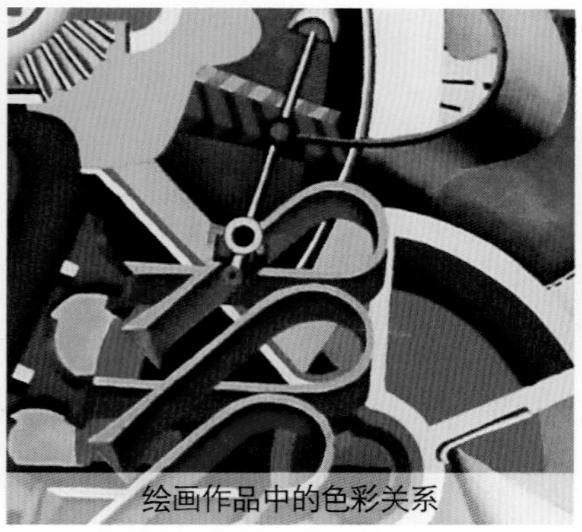

绘画作品中的色彩关系

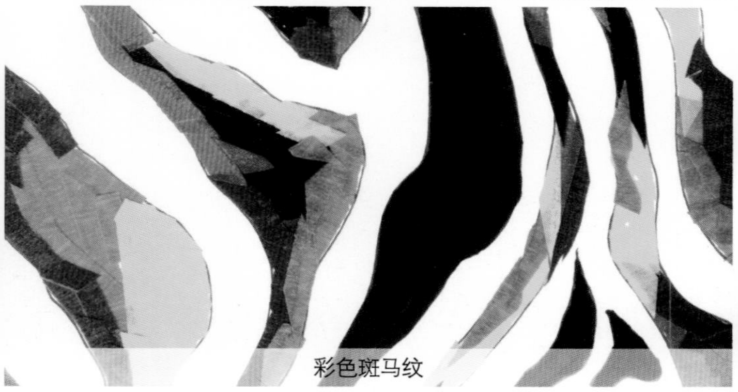

彩色斑马纹

字母墙

设计灵感

在儿童双肩背包上采用布满变形动物印花为包袋注入活力，纯色的拼接块面也让背包变得更加时尚。

Visual effects of the fusion are very strong, so we use pure color patches in kid's backpacks to make them fashionable.

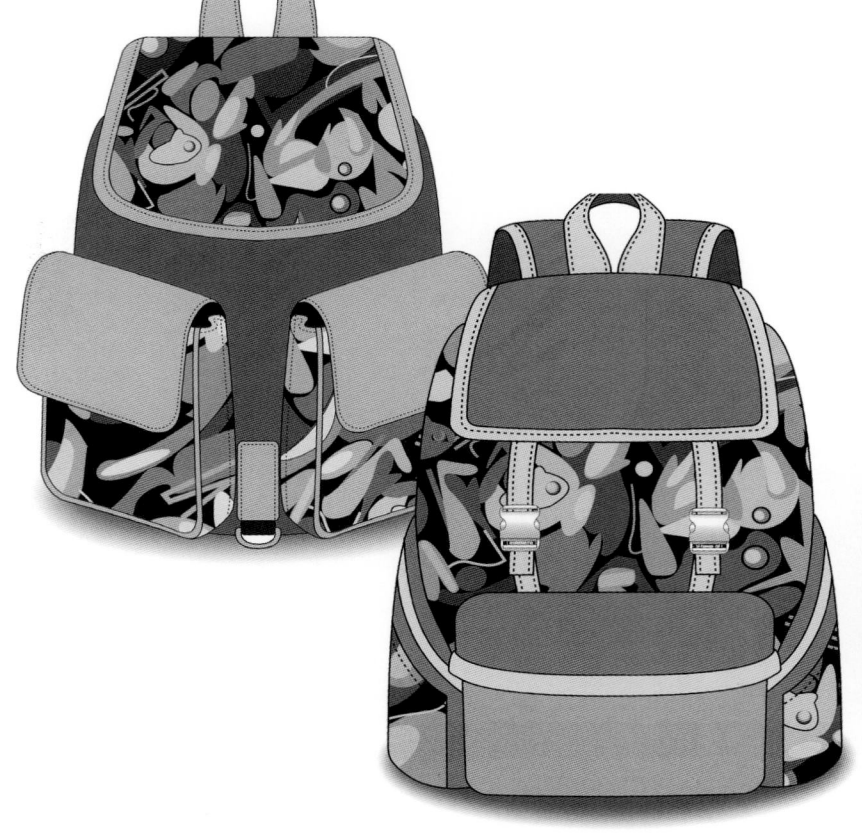

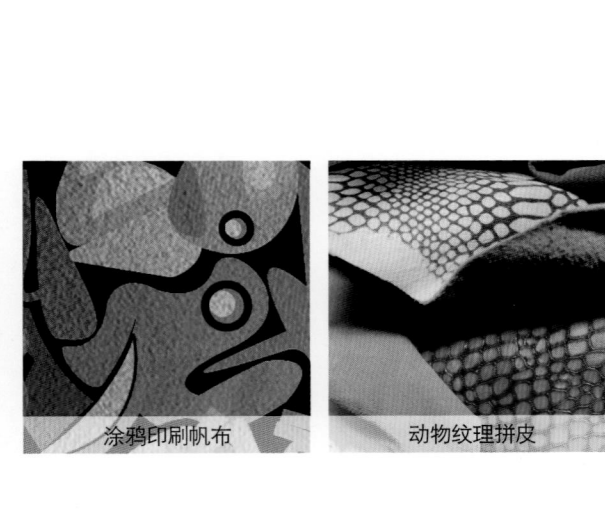

涂鸦印刷帆布

动物纹理拼皮

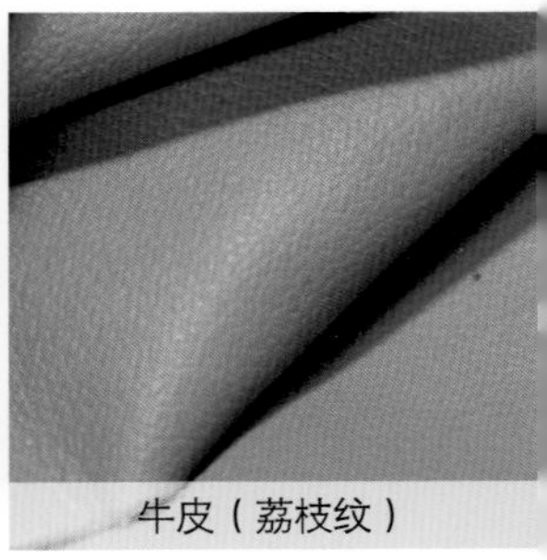

牛皮（荔枝纹）

126 www.style.sh.cn

2015 SPRING/SUMMER STYLE SHANGHAI FASHION TREND / **BAG&SUITCASE**

关 键 要 素 │ 都市街头
KEY POINTS │ CITY STREETS

幻彩红

深桃红

冰山蓝

浅草绿

迎春黄

运动桶包款式简洁，选用比较醒目的颜色，可让运动感更加鲜明
This bright color gym duffel bag is very recognizable.

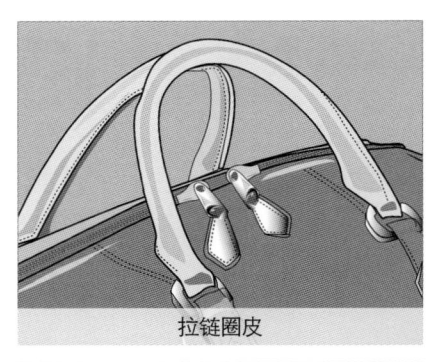

拉链圈皮

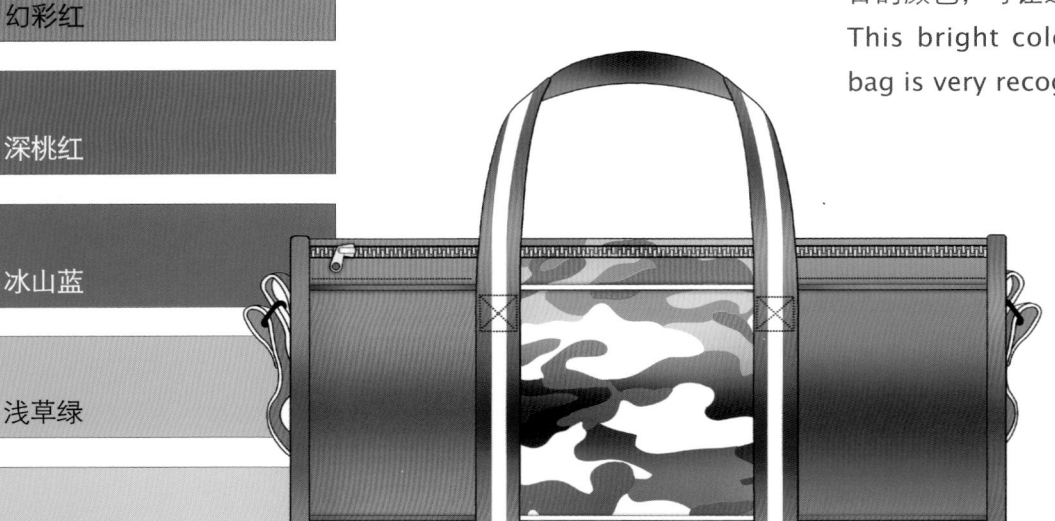

塑料插扣固定

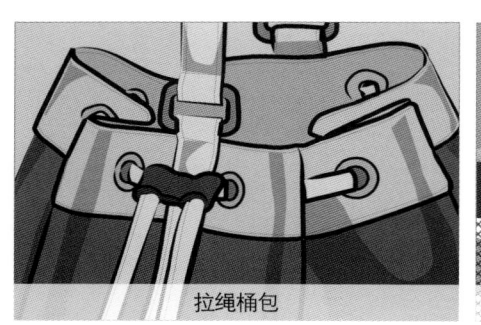

拉绳桶包

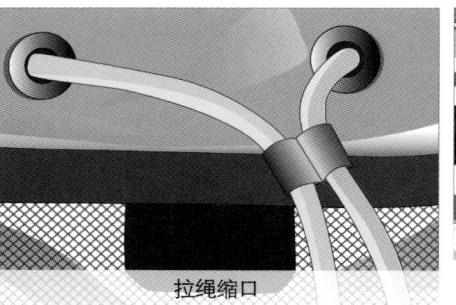

拉绳缩口

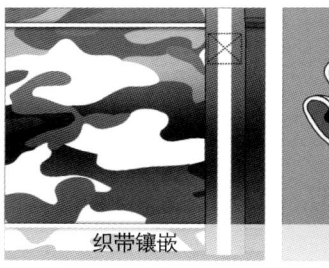

织带镶嵌

包边处理

安全拉链

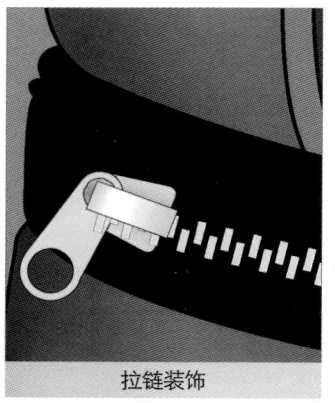

拉链装饰

网球包保持了和网球拍一样的廓型，并利用彩色迷彩来进行点缀，增强了其运动感。
This tennis bag with colorful exterior and functional structure is designed for active player.

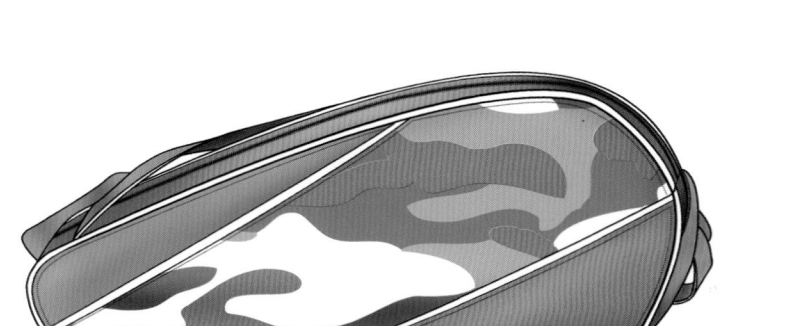

三色织带

迷彩尼龙布

牛仔布

塑料插扣

彩色拉链

127

www.style.sh.cn

帽子、首饰及其他
HAT, JEWELRY&OTHERS

海上星梦
STAR DREAMS OVER THE SEA
优活之梦
LIVING THE DREAM
时空筑梦
REALISING DREAMS THROUGH SPACE
多彩享梦
ENJOY COLORFUL DREAMS

 配饰 2015 春夏海派时尚流行趋势

海上星梦
STAR DREAMS OVER THE SEA

优雅的上海，像是一名历尽繁华，气质高贵的女子。极具地域特色的建筑线条以及繁复精致的城市元素为我们带来了无限的灵感。这样优雅与精致，高贵与创新的融合为我们的配饰品设计带来了无限灵感。

Elegant Shanghai like a noble woman who has a lot of stories. Magnificent architecture skylines with geographical features and complex city beats have brought us endless inspiration. So we bring the elegance and sophistication into our innovative hat designs.

HAT, JEWELRY & OTHERS

2015 SPRING/SUMMER STYLE SHANGHAI FASHION TREND
海派帽子、首饰及其他流行趋势

2015 SPRING/SUMMER STYLE SHANGHAI FASHION TREND / ACCESSORY

关键要素 | KEY POINTS

繁华时代
FLOURISHING ERA

精细品质
EXQUISITE QUALITY

新贵气质
NOBLE TEMPERAMENT

创新传统
INNOVATING TRADITIONS

青釉灰

青砖灰

杏仁黄

皮革黄

琥珀橙

酸枝红

紫檀棕

瓷瓦黑

铁艺灰

青瓷灰

S帽子 2015 春夏海派时尚流行趋势

海上星梦
STAR DREAMS OVER THE SEA

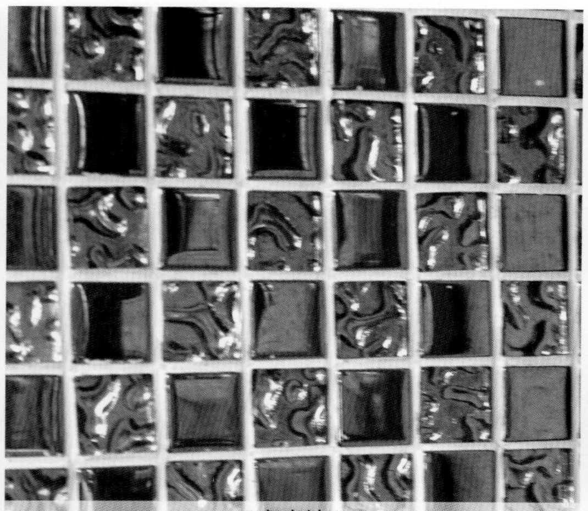

马赛克墙面 　画框边角 　金属装饰

外滩建筑灯景　装饰吊顶 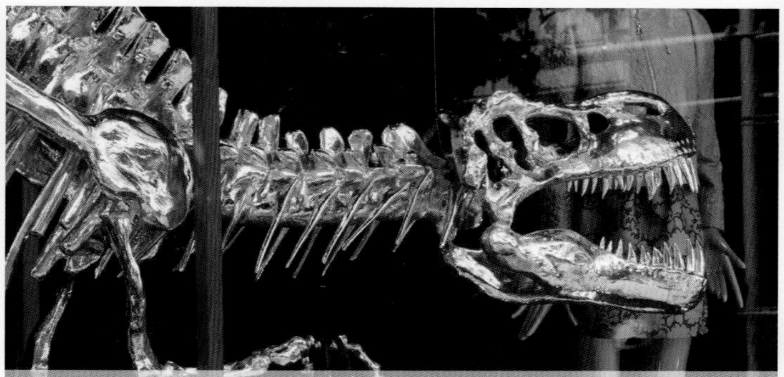　金属恐龙橱窗装饰 　创新墙面

设计灵感

传统图案的印花面料配合牛皮面料边檐上翻的款式，有种休闲感觉的繁华。

A curvy hat made of traditional printed fabrics with a spring-back brim and leather strap tops off any look with sophisticated decorations.

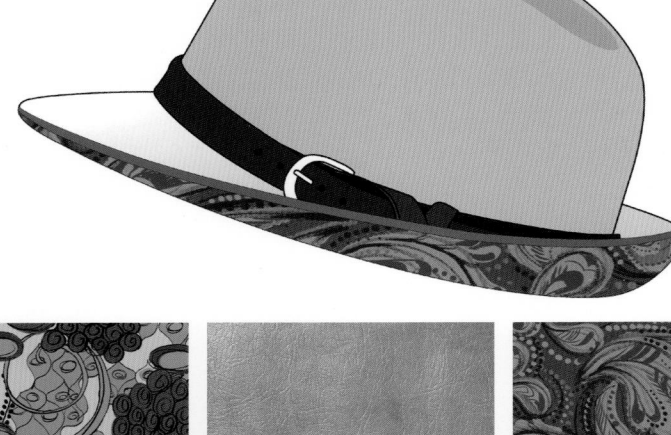

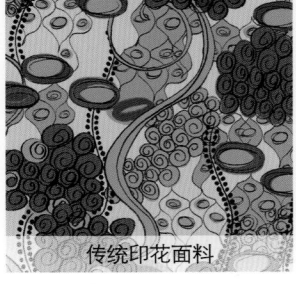

传统印花面料

牛皮

印花面料

132　www.style.sh.cn

2015 SPRING/SUMMER STYLE SHANGHAI FASHION TREND / OTHERS

关键要素 | 繁华时代
KEY POINTS | FLOURISHING ERA

皮革黄

杏仁黄

琥珀橙

酸枝红

瓷瓦黑

亚克力树脂材料的太阳眼镜，黑白参差的图形。
Resin sunglasses in black and white.

水钻配合亚克力的造型

金属缠绕式

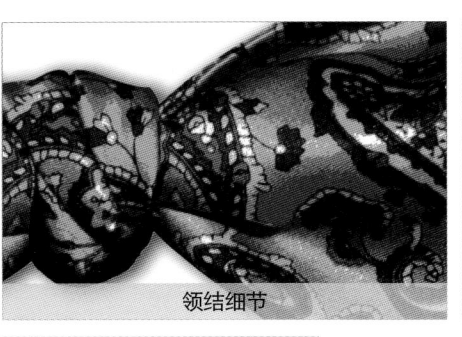

领结细节

亚克力树脂

水钻

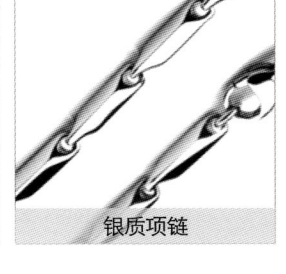

银质项链

水钻镶嵌

材料镶嵌水钻

领结作为西装的一种配饰，结合花式的印花面料，营造繁华之感。
Printed fabric together with bow as an accessory for suit makes it a luxury.

丝绸印花面料

印花面料

丝绸印花

合金

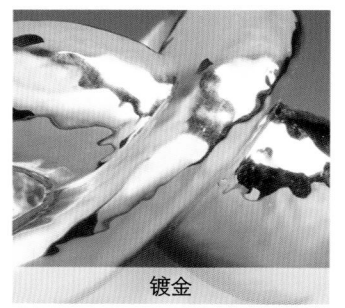

镀金

133

S 其他 2015 春夏海派时尚流行趋势

海上星梦
STAR DREAMS OVER THE SEA

帽饰品

精致餐具

玻璃天窗

琉璃杯

银制品

软饮料罐

各色食物调料

设计灵感

皮带、袖扣、假领子和项链组成精细品质的配饰，金属的皮带头，袖口上银质金属包边的贝母扣，亚克力材质或者天然的原石，金银制的细链搭配珍珠，营造精细品质的感觉。

Belt with polished buckle, sleeve buttons inlayed with mother of pearl, acrylic or natural stones, fake collar and silver or gold necklace with pearls all add charm to the whole fit.

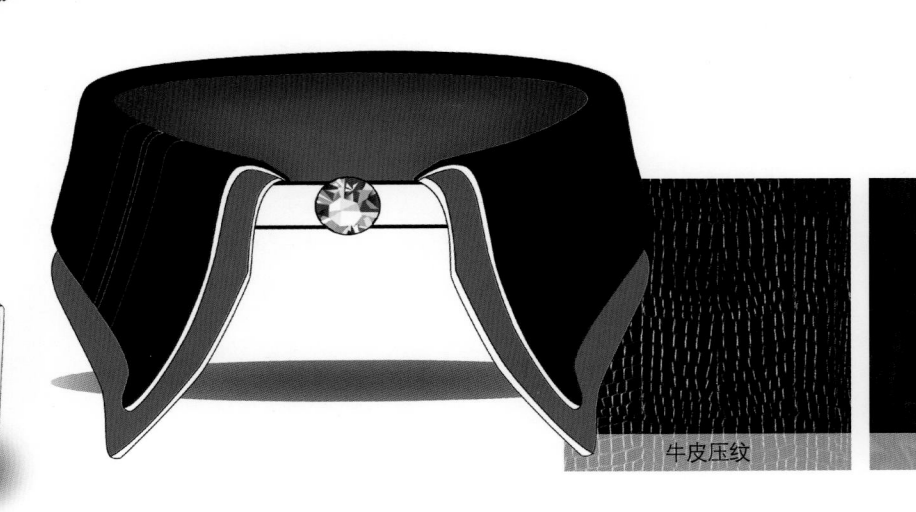

牛皮压纹

绸缎

水钻

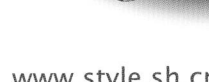

2015 SPRING/SUMMER STYLE SHANGHAI FASHION TREND / HAT

关键要素 | 精细品质
KEY POINTS | EXQUISITE QUALITY

乌梅黑

胭脂红

火焰红

浪花白

青砖灰

男士帽红黑配色，三角凹陷帽型，配上条纹和珍珠。
A black and red fedora wrapped with a band and pearls.

大帽檐

硬质帽檐

丝巾绑带

纽扣搭边

交叉条纹

V形弹性布

珍珠边条

女士帽红黑配色，和男士同色系，配上珍珠和条纹，帽檐上翻。
A brimmed black and red fedora wrapped with a band and pearls.

条纹帽边

红色里布

黑色牛皮

编织条

黑白纽扣

珍珠

S 帽子 2015 春夏海派时尚流行趋势

海上星梦
STAR DREAMS OVER THE SEA

外滩建筑

百货室内空间

欧式绘画

玻璃圆顶

拱形门雕刻纸

贵气的吊顶

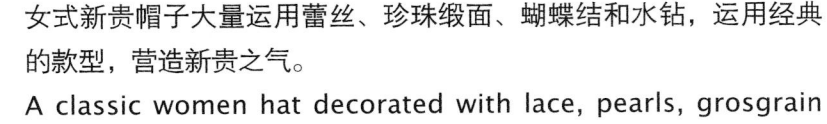
欧式铁艺

设计灵感

女式新贵帽子大量运用蕾丝、珍珠缎面、蝴蝶结和水钻，运用经典的款型，营造新贵之气。

A classic women hat decorated with lace, pearls, grosgrain band, bow tie and crystal stones.

特殊面料

竹编

水钻

136　www.style.sh.cn

2015 SPRING/SUMMER STYLE SHANGHAI FASHION TREND / OTHERS

关键要素 | 新贵气质
KEY POINTS | NOBLE TEMPERAMENT

瓷瓦黑

铁艺灰

花布蓝

浪花白

青砖灰

蓝白双层的假领子，搭配水钻的扣子和印花的面料。

Crystal stones inlayed buckle sparkles the blue and white two-layered collar cut from printed fabric.

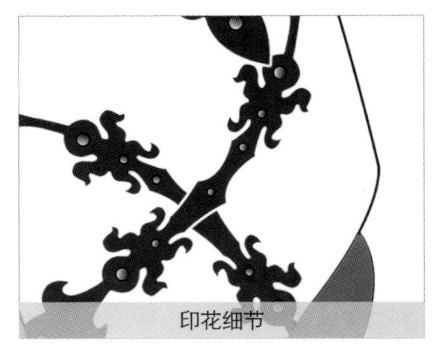

印花细节

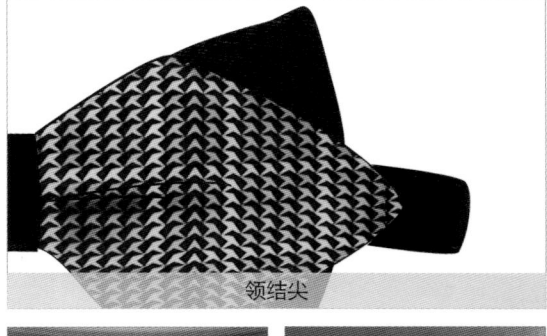

领结尖

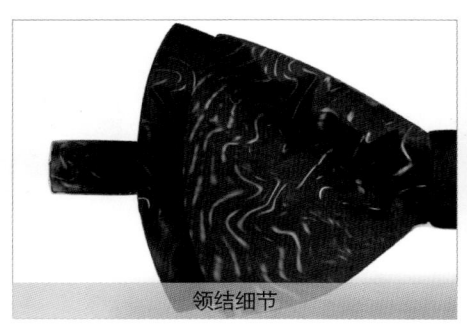

领结细节

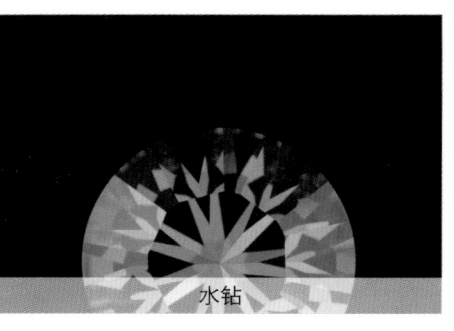

水钻

钻扣

欧式印花图案

欧式花纹

双层领结

尖形双层领结，以千鸟格和黑色绒面布料相结合。

Houndstooth prints and cotton black velvet shape a sharp two-layer bow tie with bold style.

斜格纹面料

千鸟格布料

西装面料

条纹面料

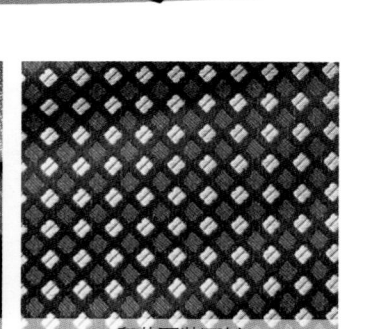

印花西装面料

137

S 首饰 2015 春夏海派时尚流行趋势

海上星梦
STAR DREAMS OVER THE SEA

橱窗纸艺展示

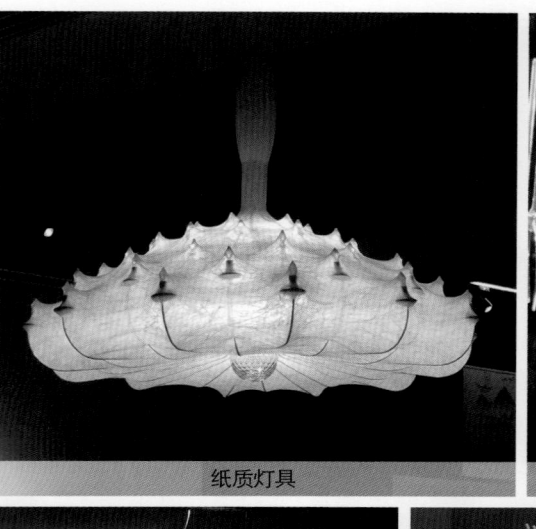

纸质灯具

金属恐龙

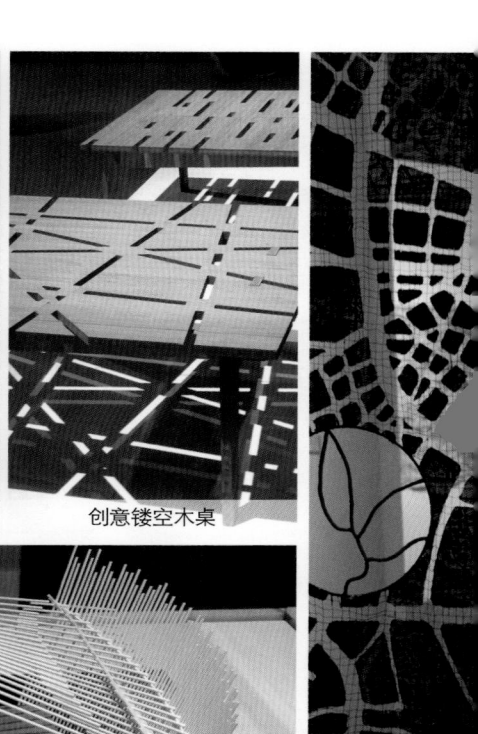

创意镂空木桌

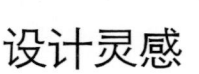

创意灯饰

塑料模型

创新剪纸

设计灵感

银质的金属链配上传统中国结的造型，运用传统元素打造的创新项链和耳环。

Silver tone chains shape in Chinese knot. Use traditional elements to create innovative necklace and earrings.

银链子

水钻镶嵌的银饰

粗银链

138 www.style.sh.cn

2015 SPRING/SUMMER STYLE SHANGHAI FASHION TREND / HAT

关键要素 | 创新传统
KEY POINTS | INNOVATED TRADITIONS

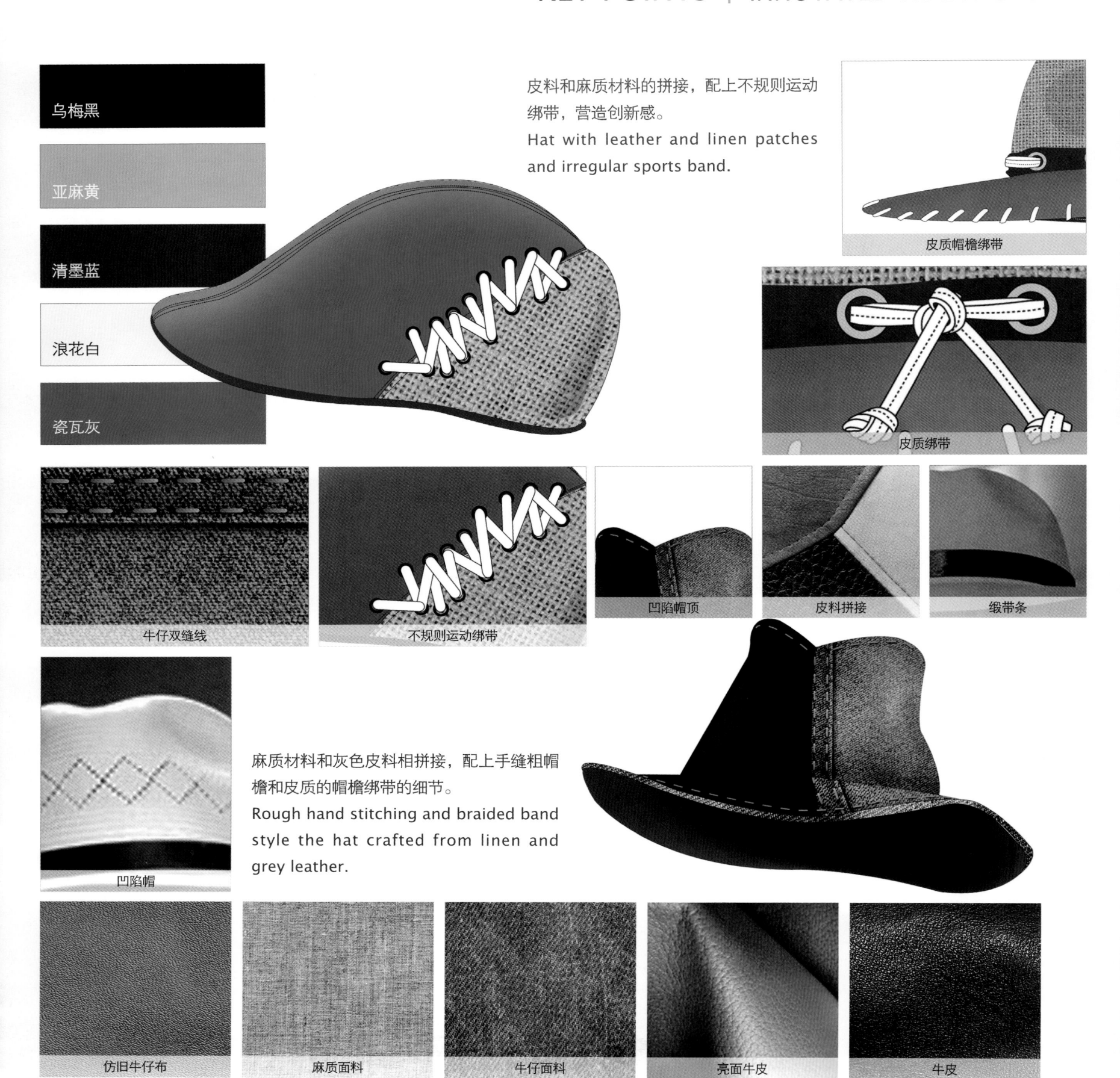

乌梅黑

亚麻黄

清墨蓝

浪花白

瓷瓦灰

皮料和麻质材料的拼接，配上不规则运动绑带，营造创新感。
Hat with leather and linen patches and irregular sports band.

皮质帽檐绑带

皮质绑带

牛仔双缝线

不规则运动绑带

凹陷帽顶

皮料拼接

缎带条

麻质材料和灰色皮料相拼接，配上手缝粗帽檐和皮质的帽檐绑带的细节。
Rough hand stitching and braided band style the hat crafted from linen and grey leather.

凹陷帽

仿旧牛仔布

麻质面料

牛仔面料

亮面牛皮

牛皮

配饰 2015 春夏海派时尚流行趋势

优活之梦
LIVING THE DREAM

上海是国际性大都市，竞争激烈而又充满挑战。在强大的压力下，人们仍然追求生活的和谐与舒适，渴望一种身心的完美平衡。在这座和谐的大都市中，人们正以健康、快乐、环保、可持续为核心理念的生活方式幸福的生活着。

As an international metropolis, Shanghai is full of competitions and challenges. Under the fierce pressure, people still pursue harmonious and comfortable life and balance of body and mind. People are live in the health, happy, environmental, sustainable way in this harmonious city.

HAT, JEWELRY & OTHERS

2015 SPRING/SUMMER STYLE SHANGHAI FASHION TREND
海派帽子、首饰及其他流行趋势

140　www.style.sh.cn

2015 SPRING/SUMMER STYLE SHANGHAI FASHION TREND / ACCESSORY

关键要素 | KEY POINTS

织补风尚
BACK TO THE BASICS

平民贵族
ORDINARY NOBILITY

生态材质
ECO MATERIAL

家庭手工
FAMILY HANDICRAFTS

淡奶黄

抹茶绿

苔藓绿

曜石黑

青花蓝

孔雀蓝

水晶蓝

豆浆白

香芋灰

紫砂棕

帽子 2015 春夏海派时尚流行趋势

优活之梦
LIVING THE DREAM

破旧的煤油灯

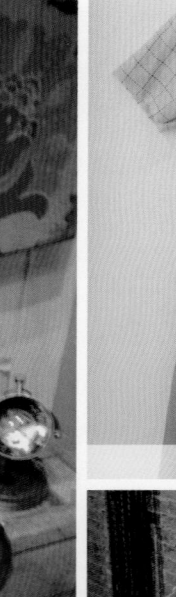

废纸群

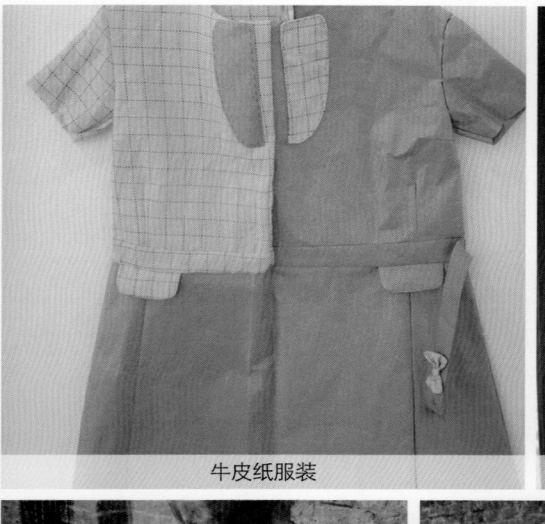

牛皮纸服装

废旧布墙面装饰

沧桑的电话亭

斑驳的墙角

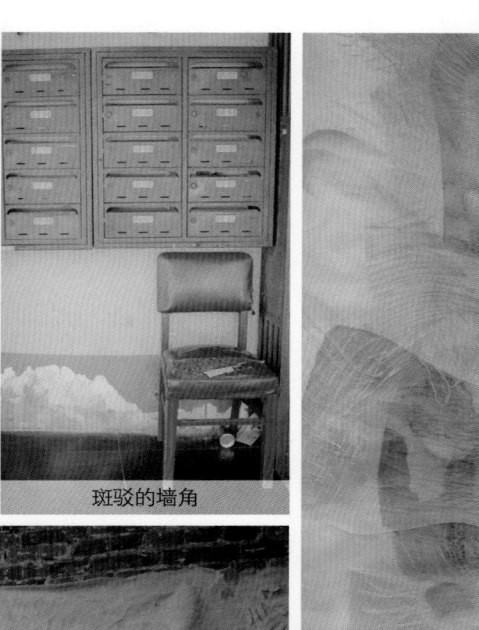

旧纱布拼接的装

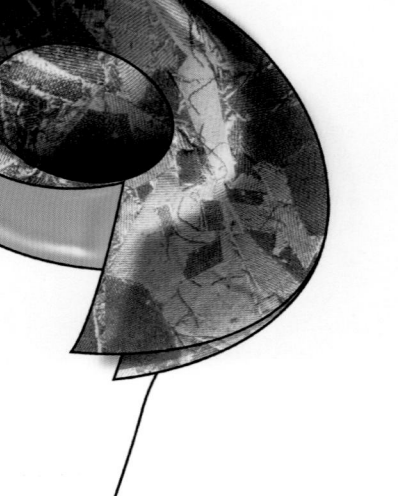

设计灵感

旧麻布、废牛仔布等拼接在一起，使原本废弃的资源再次绽放光彩。旧布料的拼接也给帽子营造一种复古的感觉。
Unused linen and denim cloth can be stitched together and be added to hat making vintage style.

旧牛仔拼布

生态材质

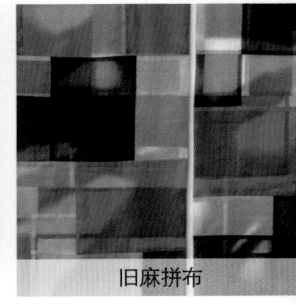

旧麻拼布

142 www.style.sh.cn

2015 SPRING/SUMMER STYLE SHANGHAI FASHION TREND / JELELRY

关键要素 | 织补风尚
KEY POINTS | BACK TO THE BASICS

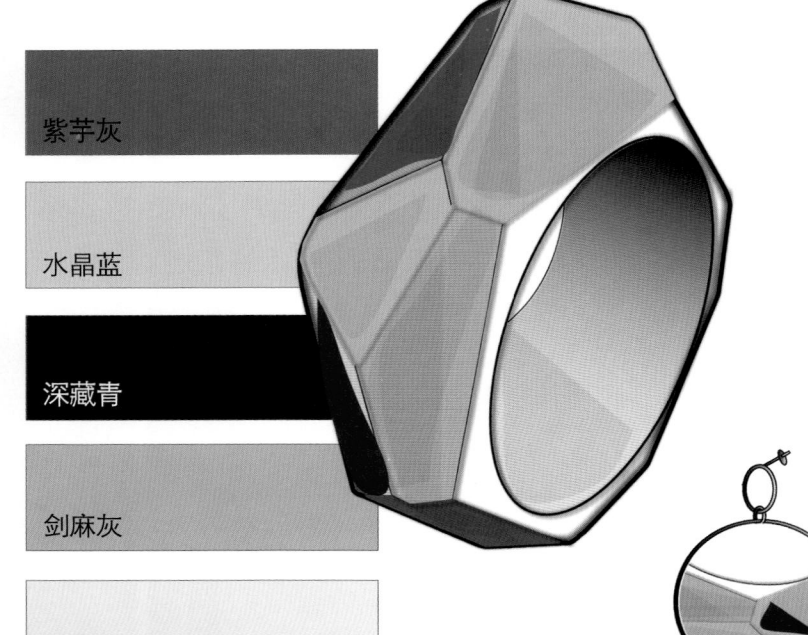

紫芋灰

水晶蓝

深藏青

剑麻灰

豆浆白

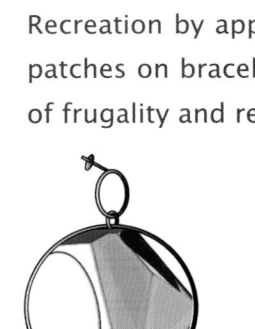

节俭设计首饰意图体现节约和循环利用的设计概念。

Recreation by applying wasted fabric patches on bracelet conveys the idea of frugality and re-use.

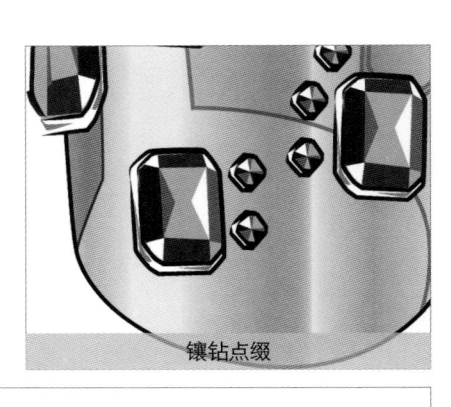

镶钻点缀

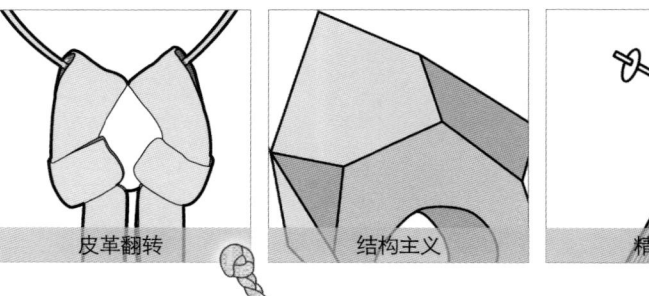

太空铝合金

菱角切割

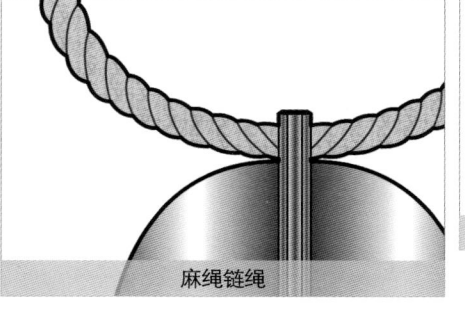

麻绳链绳

皮革翻转

结构主义

精巧耳钉

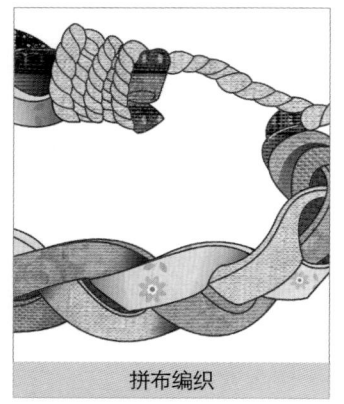

拼布编织

家庭手工艺的编织美感和细节大量的运用到绑带和装饰细节，体现传统手工艺的精致与细腻。

Traditional handicrafts have a lot of weaving and decorative details, such as bamboo weaving, knitting, crochet knitting, fishbone braid, tightly woven multi-strand knitting, classic knitting, etc. A wide variety of knitting patterns bring more collisions and styles.

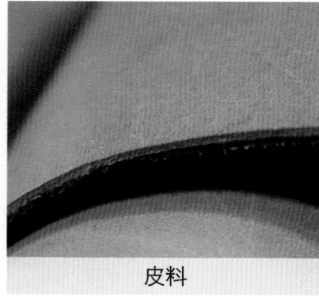

皮料

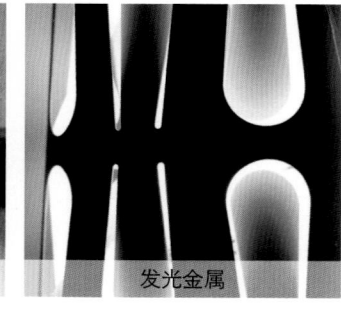

发光金属

亚克力

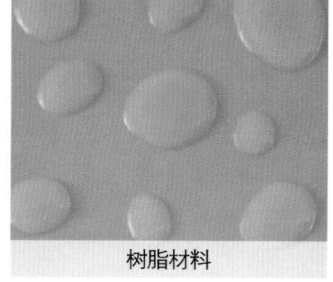

树脂材料

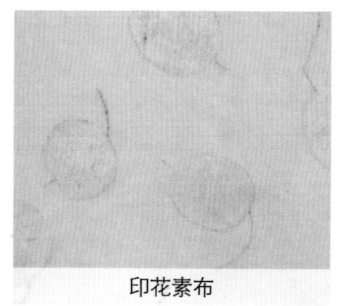

印花素布

143

其他

2015 春夏海派时尚流行趋势

优活之梦
LIVING THE DREAM

平民手工艺者

精致餐厅一角

复古皮沙发

贵妇油画像

精致餐具

中国围棋

三毛卖报场最

设计灵感

这系列腕表设计简洁大方。普通的材质加上独特的贵族纹样低调中透露着些许奢华。
Simple but functional design of this collection together with unique prints
makes the watch a piece of luxury.

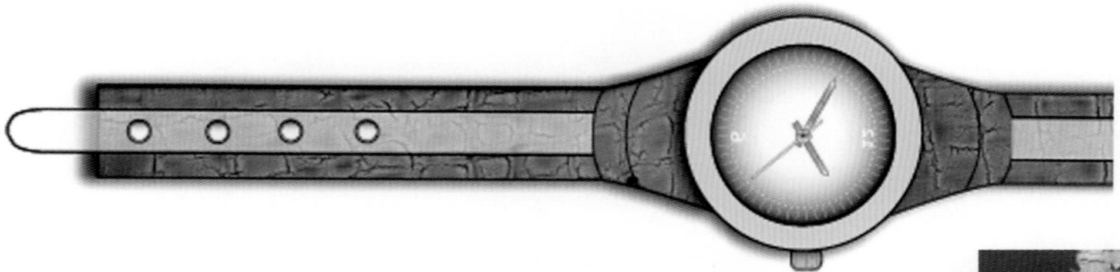

钻石

亮面亚克力

树脂材料

144　www.style.sh.cn

2015 SPRING/SUMMER STYLE SHANGHAI FASHION TREND / HAT

关键要素 | 平民贵族
KEY POINTS | ORDINARY NOBILITY

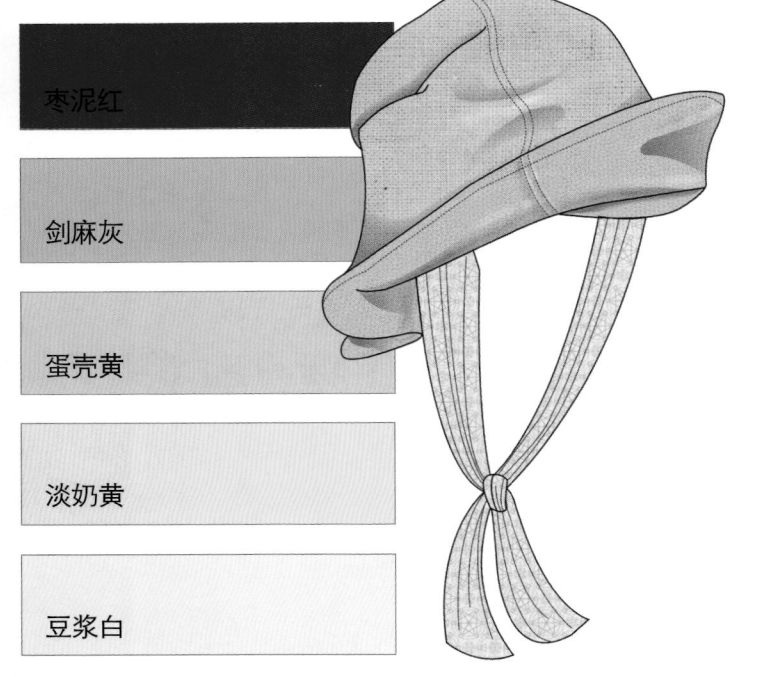

枣泥红
剑麻灰
蛋壳黄
淡奶黄
豆浆白

平民也能拥有一个贵族梦，平民贵族帽子多运用平民化的面料搭配精致的皮料。
Cheap fabrics together with well-textured leather can also look expensive.

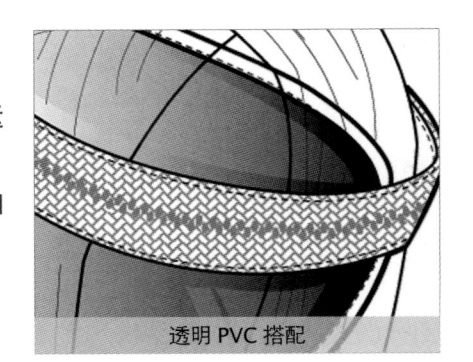
透明 PVC 搭配

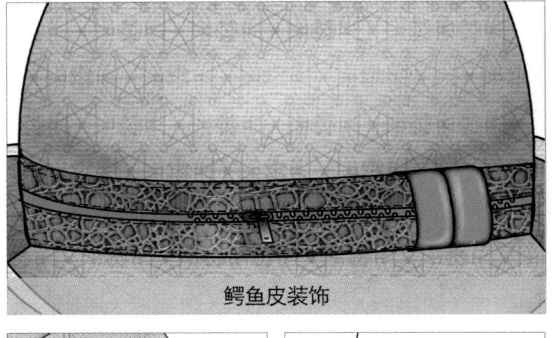
鳄鱼皮装饰

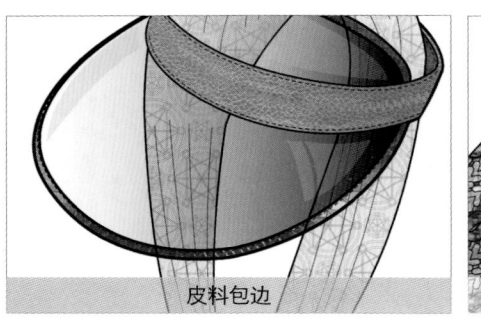
皮料包边

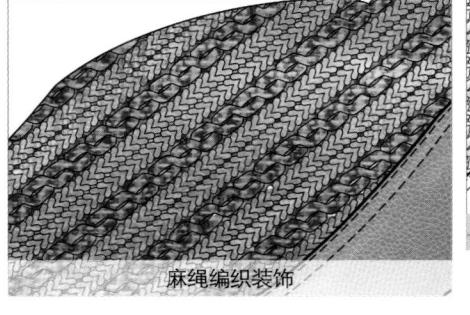
麻绳编织装饰

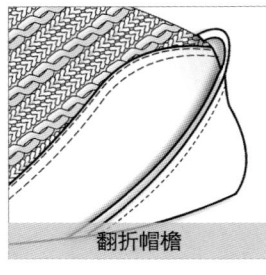
翻折帽檐

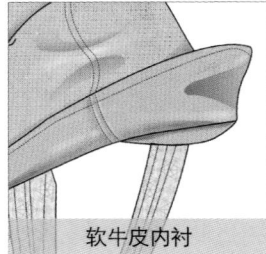
软牛皮内衬

翻折边缘

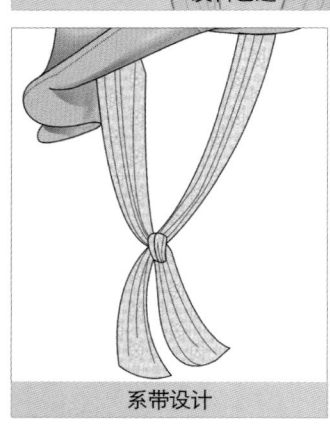
系带设计

礼帽风格的帽型彰显贵族气息，舒适的麻布材料让帽子糅合了些许自然的味道。拉链的点缀也可圈可点。
A natural linen texture marks a vintage-style trilby with a zipper.

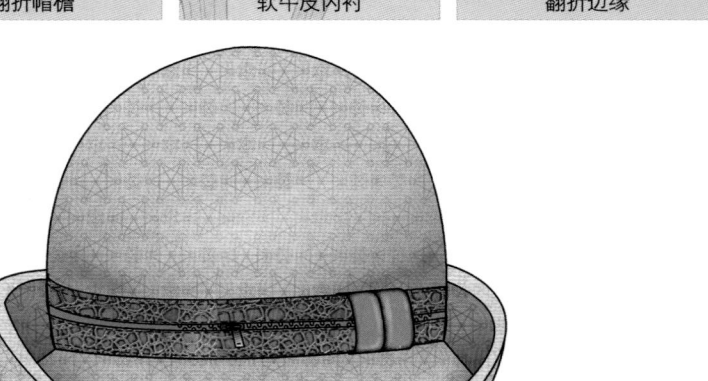

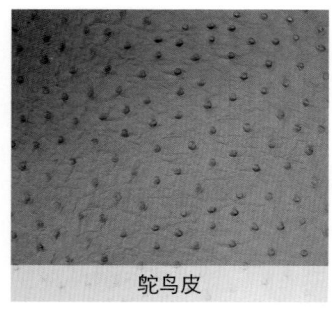
驼鸟皮

条纹印花麻布

亚麻布

皮料

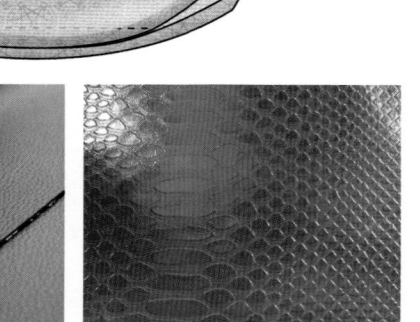
鳄鱼皮

S帽子 2015 春夏海派时尚流行趋势

优活之梦
LIVING THE DREAM

竹装置艺术

竹编制装饰

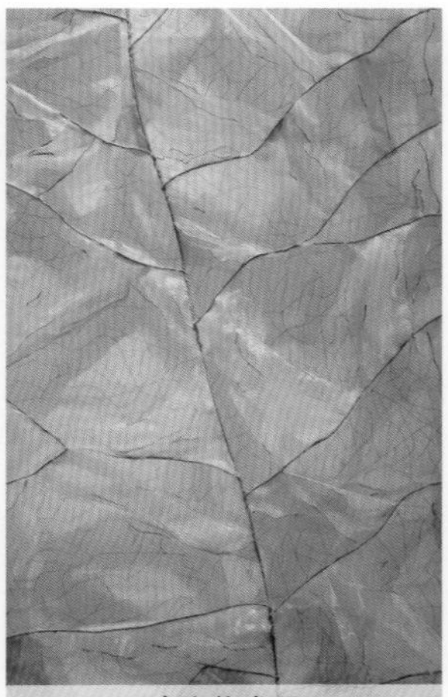

叶脉艺术品

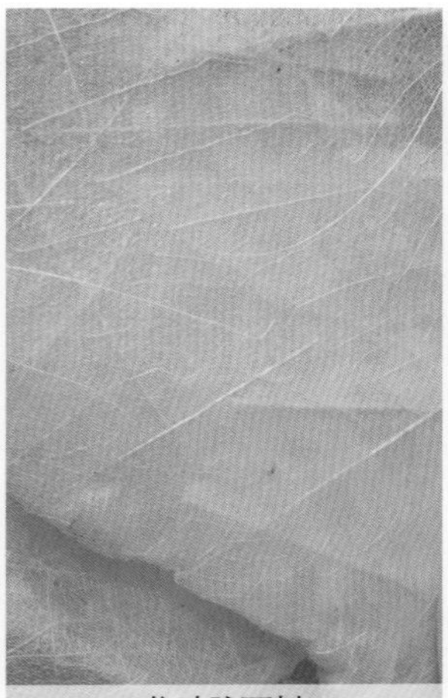

仿叶脉面料

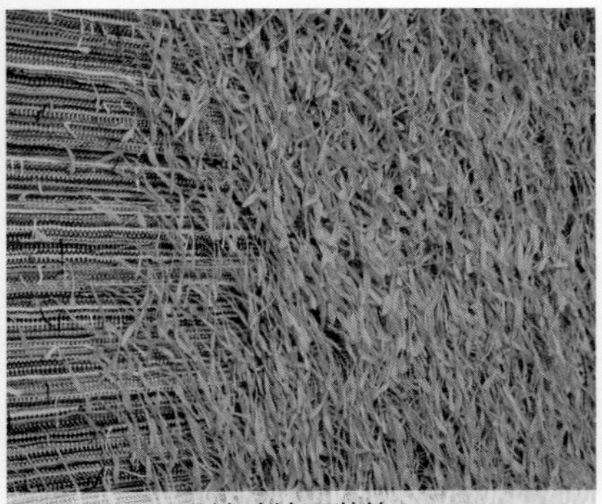

枯叶墙面装饰

竹材质自行车

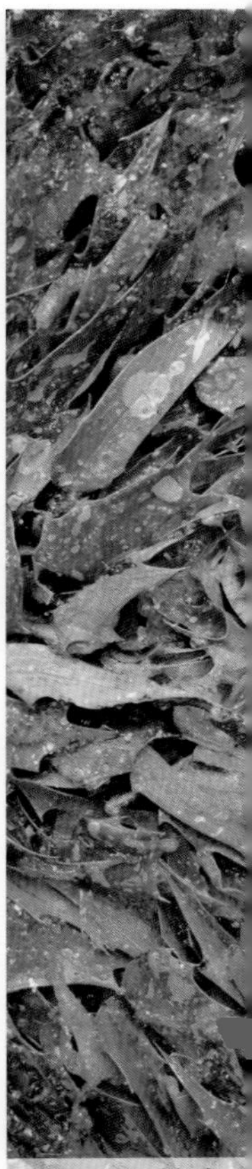

枯叶

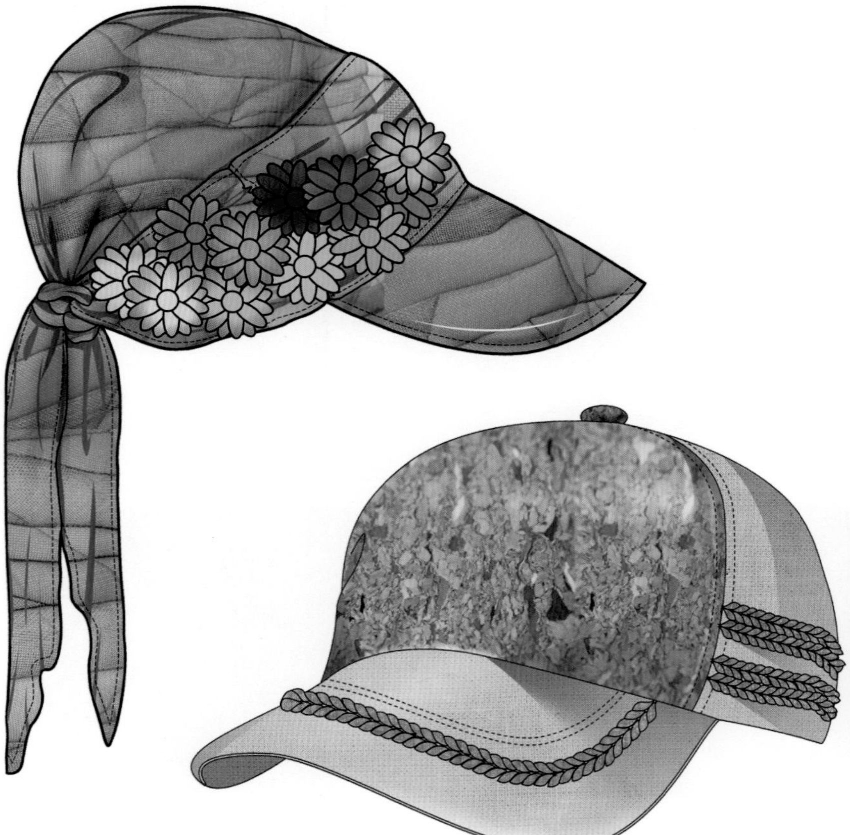

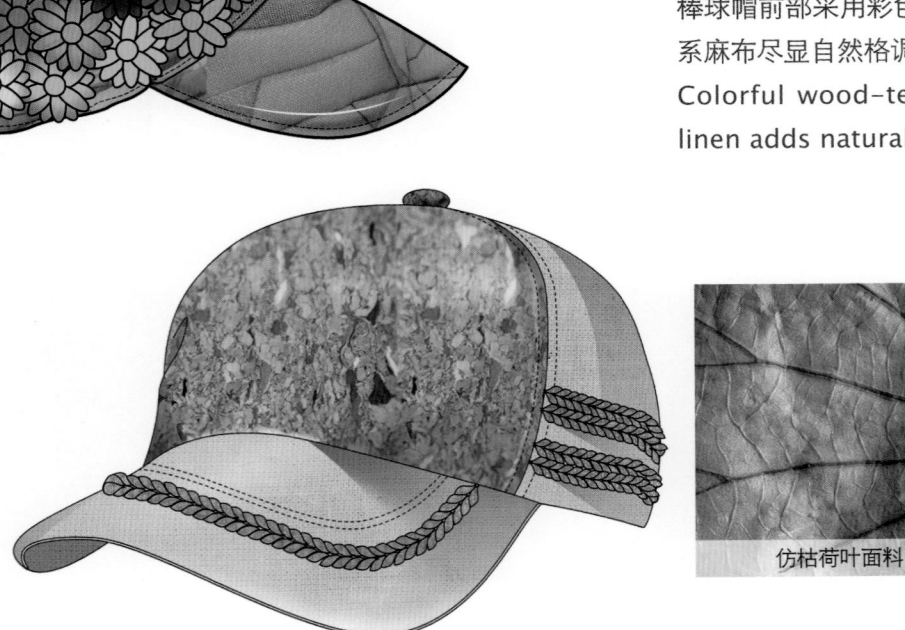

设计灵感

棒球帽前部采用彩色软木材料，自然之中又充满了些许活泼。配上同色系麻布尽显自然格调。麻绳的点缀也很好呼应了生态材质这个主题。
Colorful wood-texture front panel together with similar tone linen adds natural style to this cap.

仿枯荷叶面料

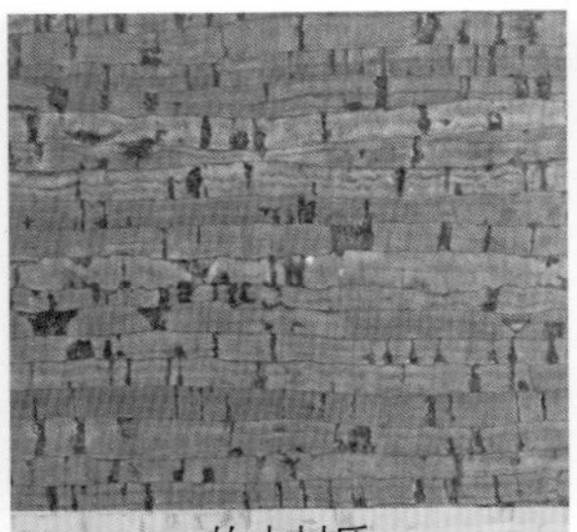

软木材质

蒲公英花图案面料

146 www.style.sh.cn

2015 SPRING/SUMMER STYLE SHANGHAI FASHION TREND / **OTHERS**

关键要素 | 生态材质
KEY POINTS | ECO MATERIAL

青花蓝

紫砂棕

孔雀蓝

淡奶黄

艾草绿

自然肌理以琥珀的形式呈现在戒指和项链等原放珠宝的位置，肌理感增强。
Use ambers instead of rhinestones on rings and necklaces.

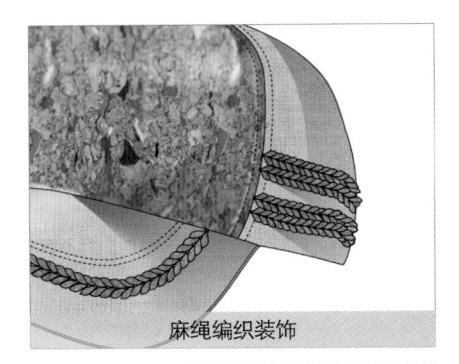

麻绳编织装饰

面料塑造自然花卉形态

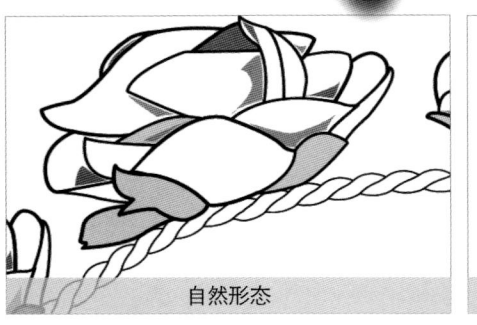

自然形态

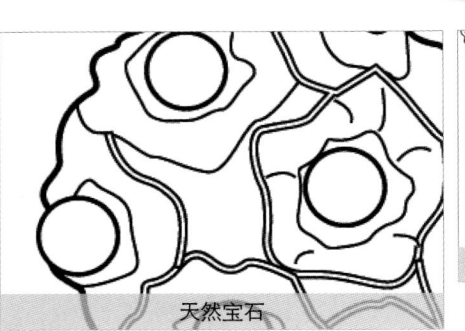

天然宝石

生态镂空

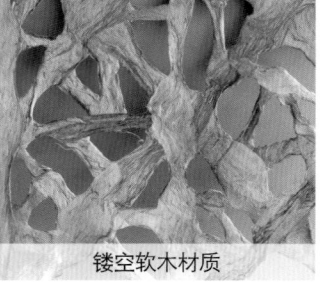

鸭舌帽檐

软木肌理

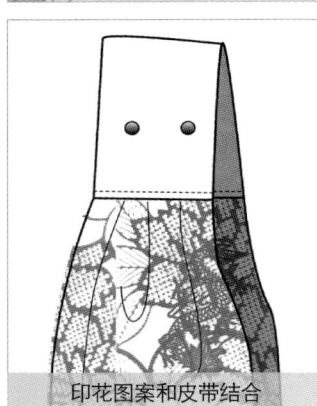

印花图案和皮带结合

墨镜太阳镜是近些年无论T台还是街拍的时尚必备单品。在纯色镜框上运用生态材质图案，呈现自然生态质感。
Sunglasses are essentials both on T stage and street style. Nature texture adds vitality to pure color frame.

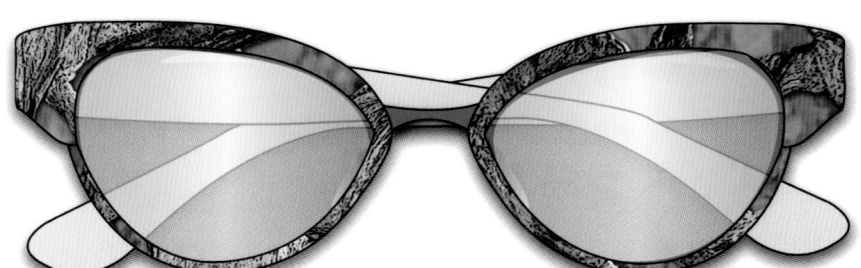

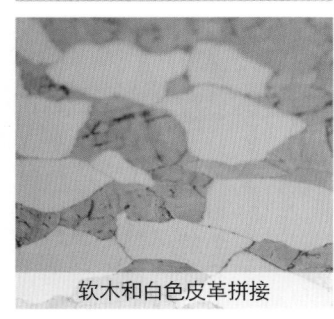

软木和白色皮革拼接

仿镂空树皮

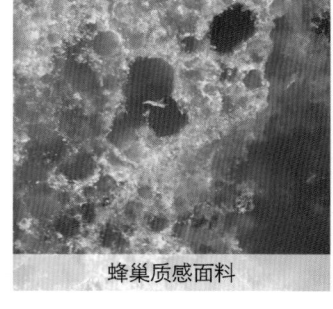

蜂巢质感面料

镂空软木材质

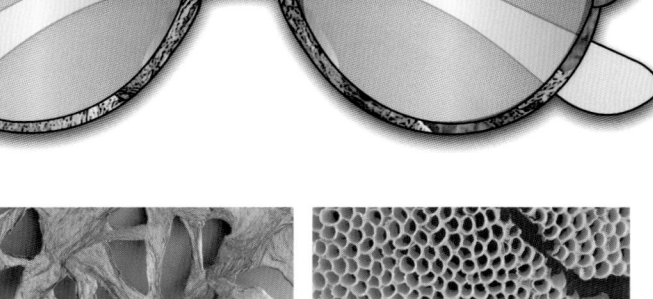

蜂窝聚合图案离型纸

S帽子 2015 春夏海派时尚流行趋势

优活之梦
LIVING THE DREAM

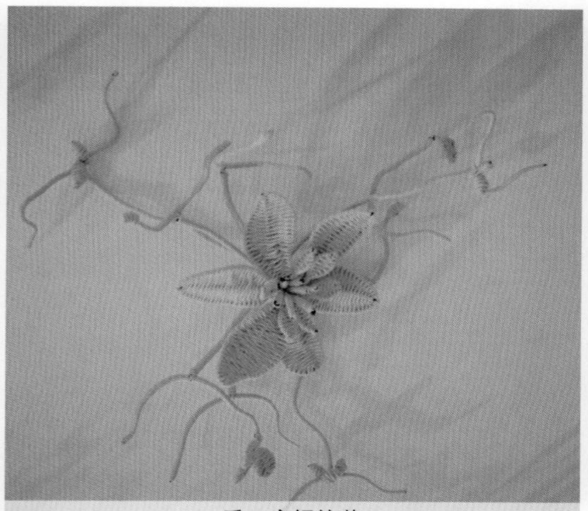

手工麻绳编花

手工扣子装饰

手工缝制工具

彩色手工编织装饰

工具箱

铁艺编织灯罩

手工艺品墙面装

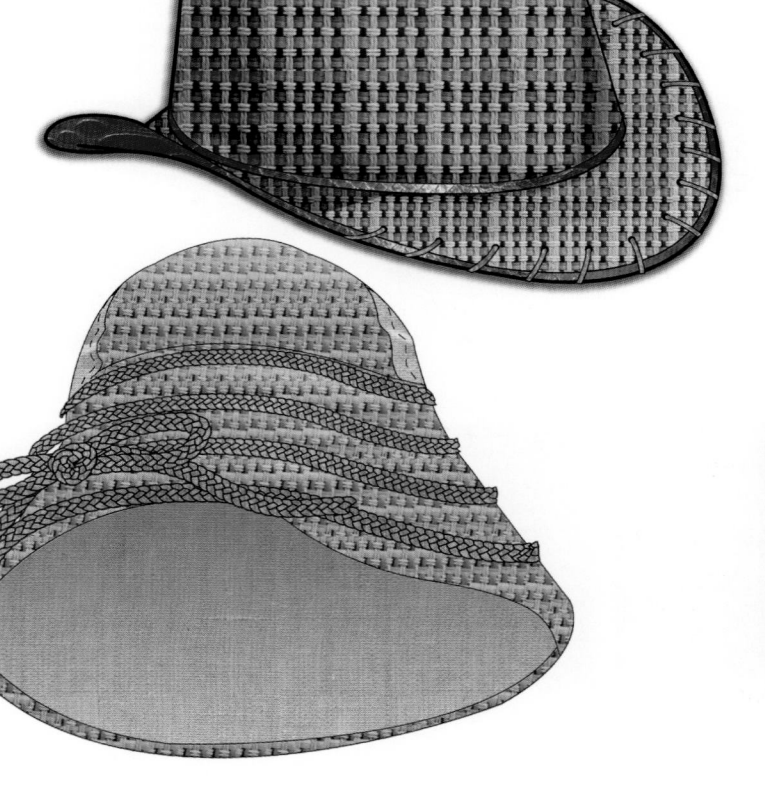

设计灵感

牛仔帽大面积运用了编织材料。在编织材料中加入各种缤纷的色彩，让这一季的春夏不再沉闷。尽情追求阳光、沙滩和海浪吧。

Weaving material dyed with different colors forms a passionate appeal on the cowboy hat and the fedora. Wear them and enjoy your spring and summer on a beach.

麻绳编织面料

综合材料编织面料

皮料编织面料

148 www.style.sh.cn

2015 SPRING/SUMMER STYLE SHANGHAI FASHION TREND / **OTHERS**

关 键 要 素 | 家庭手工
KEY POINTS | FAMILY HANDICRAFTS

编织条带替换原有单一面料，呈现特殊韵味。
Use braided fabric to replace single material.

麻花编织表带

双绳子缠绕绑带装饰

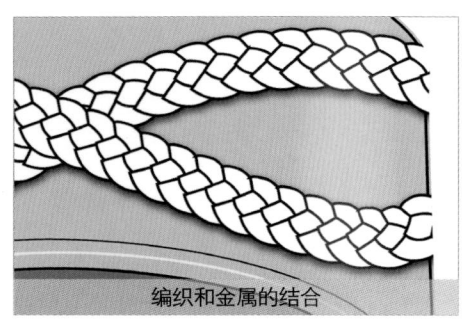
编织和金属的结合

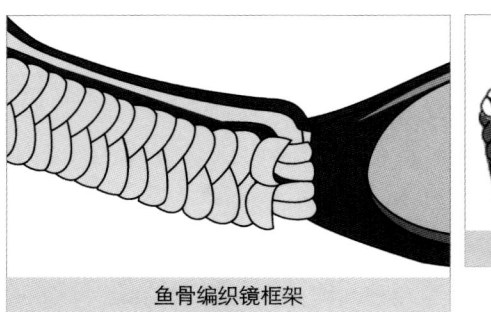
鱼骨编织镜框架

钩针细节

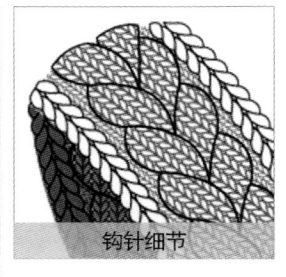
多股线绳

彩色多重编织

编织腕带

编织花式种类繁多，以图案的方式呈现在塑料、软木等其他材质上进而运用到配饰上，呈现出家庭手工艺的精细质感。
Differently braided prints on plastic or wood jewelries make you think of handicrafts with good quality.

软木和草编复合编织材料

彩色竹编

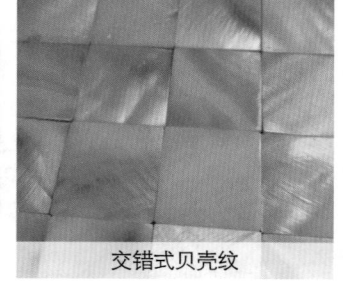
交错式贝壳纹

塑料编织

树皮拼接

孔雀蓝

紫砂棕

水晶蓝

豆浆白

香芋灰

149

 配饰 2015 春夏海派时尚流行趋势

时空筑梦
REALISING DREAMS THROUGH SPASE

"上海中心"建筑外观呈螺旋式上升，建筑表面的开口由底部旋转贯穿至顶部，犹如城市天际线。而日新月异发展的陆家嘴就像一只巨大的金角兽伸出脑袋张开嘴巴在浦江之东畅饮。

"Shanghai Tower" takes the form of nine cylindrical buildings stacked atop each other spirally and a notch along the building facade rotates from the bottom to the top. It symbolizes the dynamic emergence of the fast—developing modern Lujiazui located along the Huangpu River.

HAT, JEWELRY & OTHERS

2015 SPRING/SUMMER STYLE SHANGHAI FASHION TREND
海派帽子、首饰及其他流行趋势

150　www.style.sh.cn

2015 SPRING/SUMMER STYLE SHANGHAI FASHION TREND / **ACCESSORY**

关键要素 | KEY POINTS

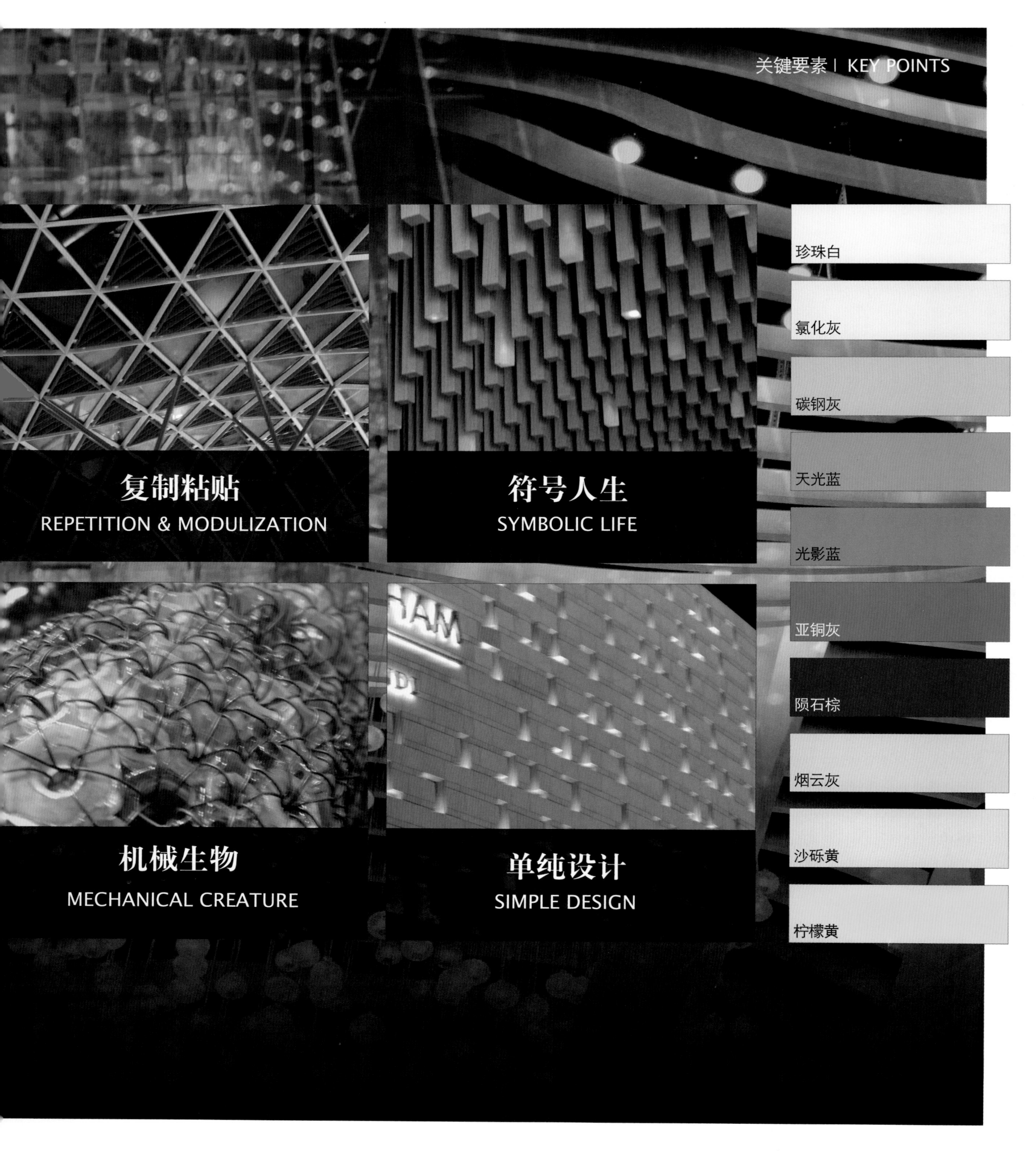

复制粘贴
REPETITION & MODULIZATION

符号人生
SYMBOLIC LIFE

机械生物
MECHANICAL CREATURE

单纯设计
SIMPLE DESIGN

珍珠白

氯化灰

碳钢灰

天光蓝

光影蓝

亚铜灰

陨石棕

烟云灰

沙砾黄

柠檬黄

 **首饰** 2015 春夏海派时尚流行趋势

时空筑梦
REALISING DREAMS THROUGH SPACE

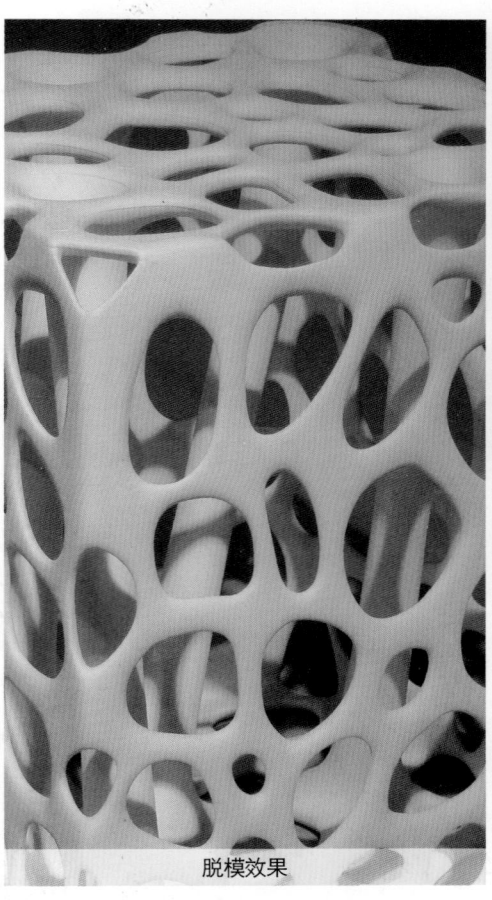

脱模效果

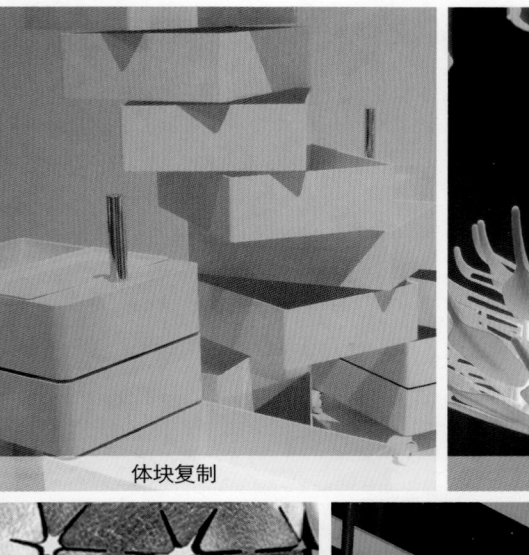

体块复制

片状结构复制

线状发散

体块复制

三角复制

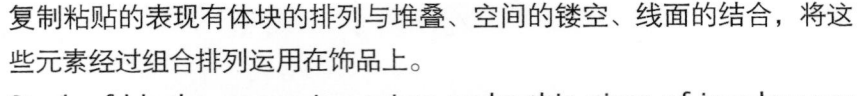

灯管排布

体块复制

设计灵感

复制粘贴的表现有体块的排列与堆叠、空间的镂空、线面的结合，将这些元素经过组合排列运用在饰品上。

Stack of blocks, engraving prints make this piece of jewelry very artistic.

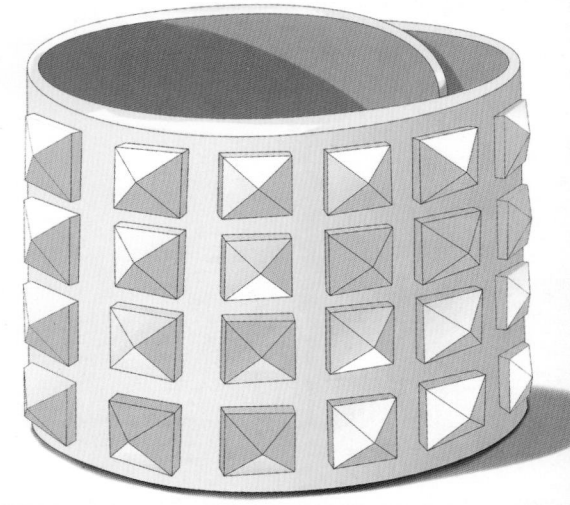

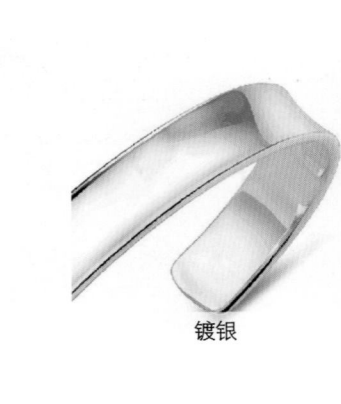
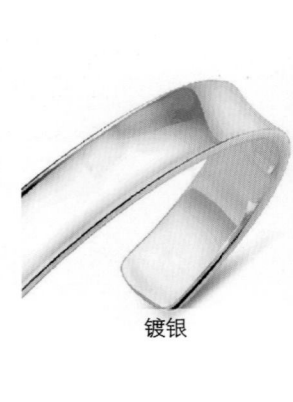
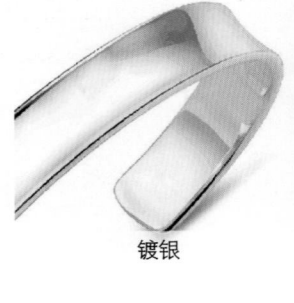

镀银

硅胶铆钉

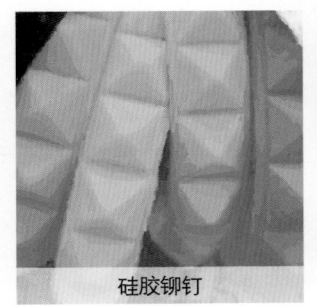

皮质手袋

152   www.style.sh.cn

2015 SPRING/SUMMER STYLE SHANGHAI FASHION TREND / HAT

关键要素 | 复制粘贴
KEY POINTS | REPETITION & MODULIZATION

光影蓝

天光蓝

珍珠白

烟云灰

碳钢灰

同一个图形有时通过等比例放大缩小会形成奇妙的效果，金属材质的运用增加了光线的反射角度。

Different sizes of same print can create unexpected effect and metallic material brightens the hat with more reflection.

网面堆叠

金属软管缠绕

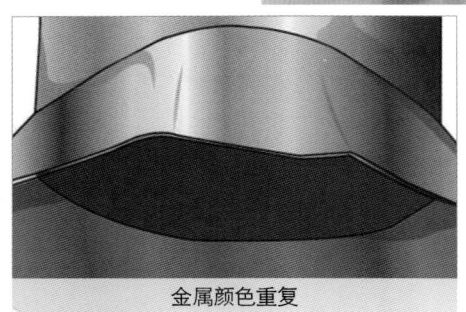

金属颜色重复

片状结构重复

晶体排列重复

扭曲重复

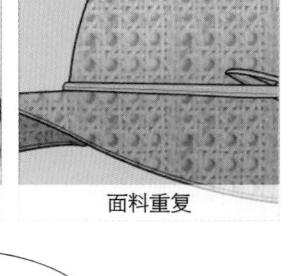

面料重复

将鞋舌复制的概念沿用至帽檐上，可以看到原本单层的帽檐在重复后变得动感活力了。

Turning single layer design into multi ones adds vitality to the hat.

条带组合重复

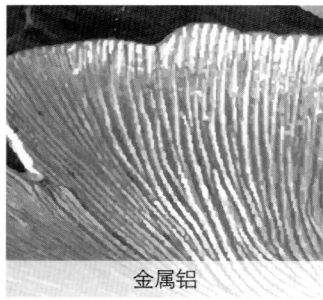

金属铝

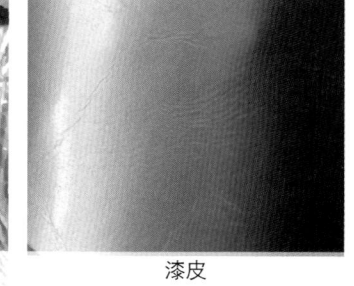

漆皮

亮彩PVC

棉

皮革编织

153

时空筑梦
REALISING DREAMS THROUGH SPACE

2015 春夏海派时尚流行趋势

鞋图组合的花瓣图

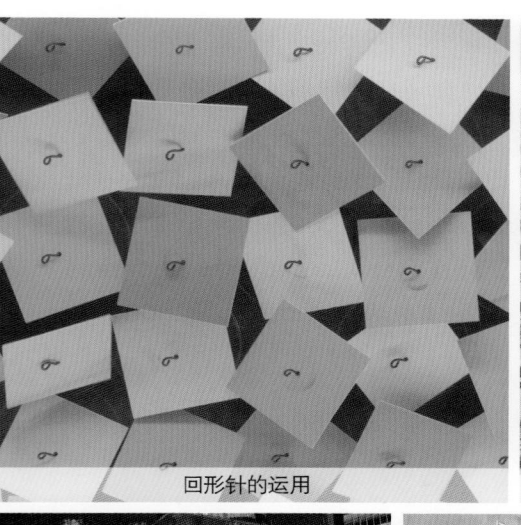

回形针的运用

抽象花瓣

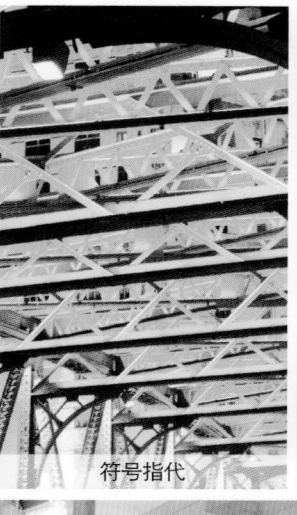

符号指代

轮廓即为符号

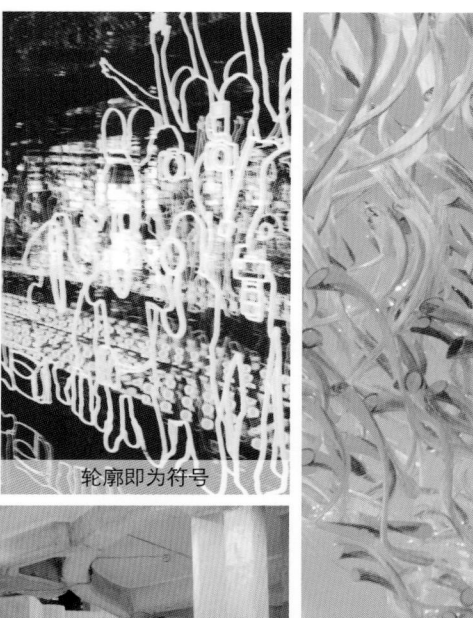

电影符号代表

颜料发散状涂

设计灵感

未来符号设计中最直接的就是符号的直接运用了。运用各式各样的符号来表现信息时代的快速与新奇。

Various signs and symbols stand for fast and novelty information age. We use them directly in our designs adding futuristic style.

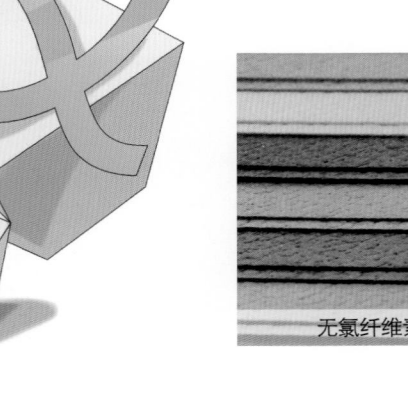

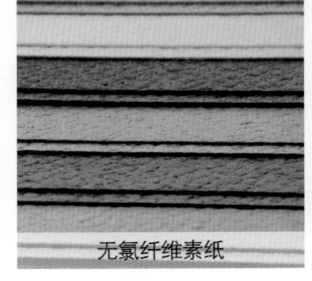

无氯纤维素纸

草编

印花面料

154   www.style.sh.cn

2015 SPRING/SUMMER STYLE SHANGHAI FASHION TREND / OTHERS

关键要素 | 符号人生
KEY POINTS | SYMBOLIC LIFE

光影蓝

天光蓝

珍珠白

苹果绿

沙砾黄

将符号图形运用在戒指上，富有个性且时尚感强。
Printing symbols on belt is stylish and fashionable.

手环上的几何符号

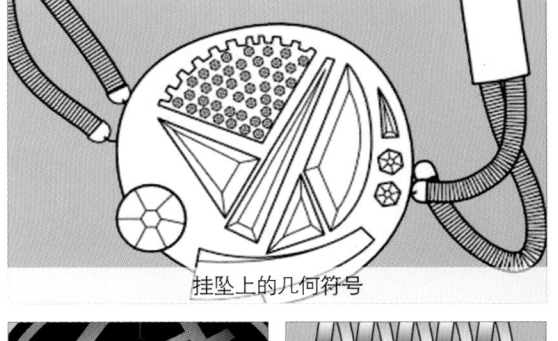

挂坠上的几何符号

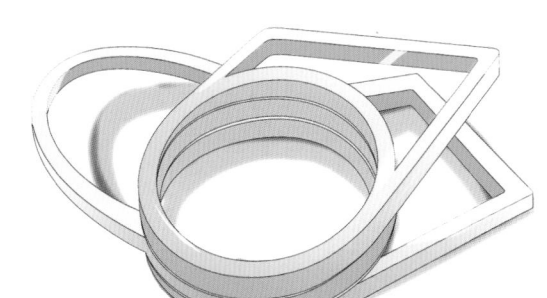

戒指的鱼眼形态

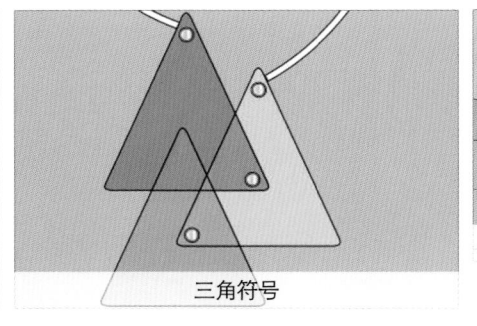

三角符号

符号贴图

手表上的地图符号

护腕上的时刻表

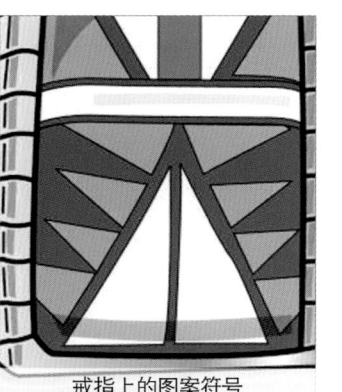

戒指上的图案符号

手机是当下人手必备的通讯工具，手机外壳的
符号图形设计更能彰显一个人的个性和趣味。
Every one carries a cell phone. Phone
cases with customized graphic design
will be very special.

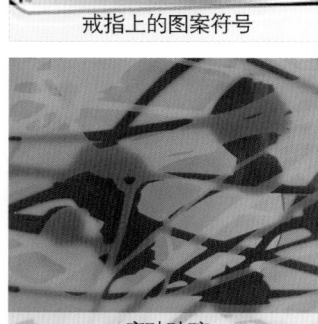

磨砂玻璃

金属铆钉

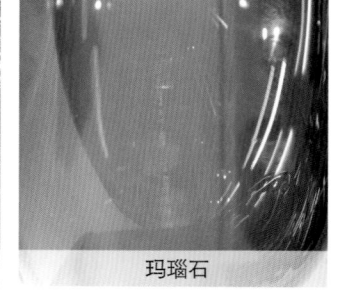

玛瑙石

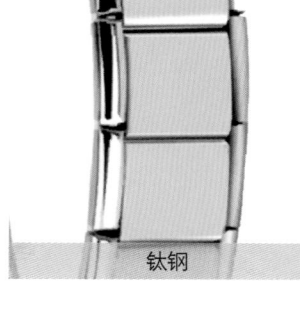

钛钢

人造毛

时空筑梦
REALISING DREAMS THROUGH SPACE

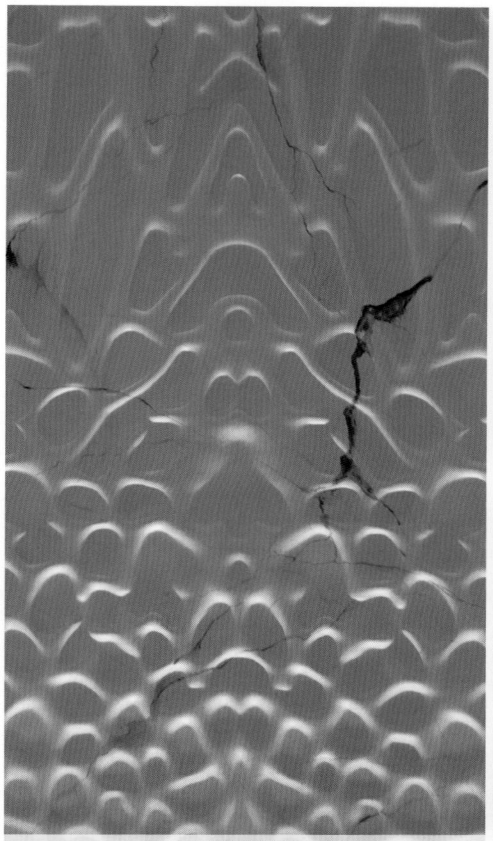

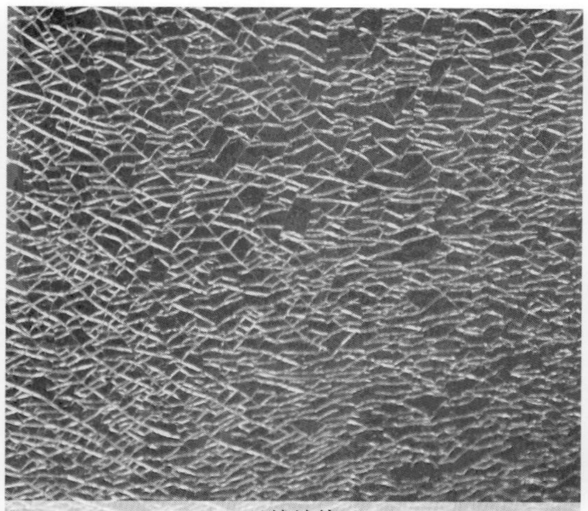

丝线缠绕

灯管

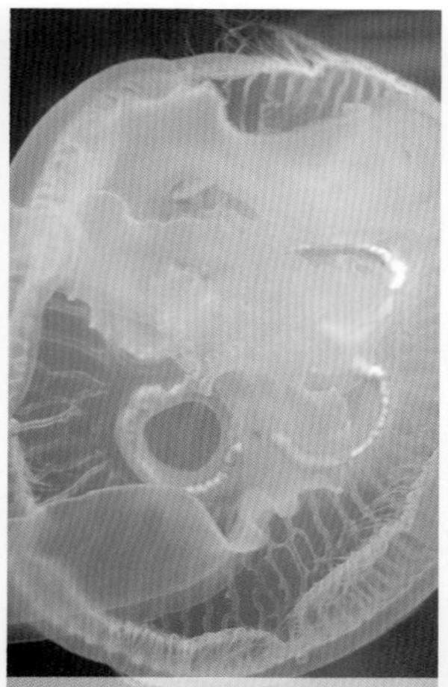

海蜇

水泡肌理

模拟火山坑

机械化仿生

光斑

设计灵感

机械女帽以水母为设计原型，通过帽檐的弯曲表现海洋生物的柔和线条。

Inspired by jellyfish, wavy brim brings feminine glamour to this hat.

数码印花面料

数码印花面料

毛毡

2015 SPRING/SUMMER STYLE SHANGHAI FASHION TREND / OTHERS

关键要素 | 机械生物
KEY POINTS | MECHANICAL CREATURE

玻璃镜面上的块面装饰，使得眼睛更具神秘感。
Patches of decoration on glasses add a sense of mystery to eyes.

抽象生物主题在饰品上的表现方式多在形态上，如仿生物造型，多线条排列组合在护腕上，使饰品更灵动富有生命。
The expression of the abstract creatures in the accessories is mainly in the shape, for example, the multi-lines design which modeling the shape of the creatures makes wristbands more vivid.

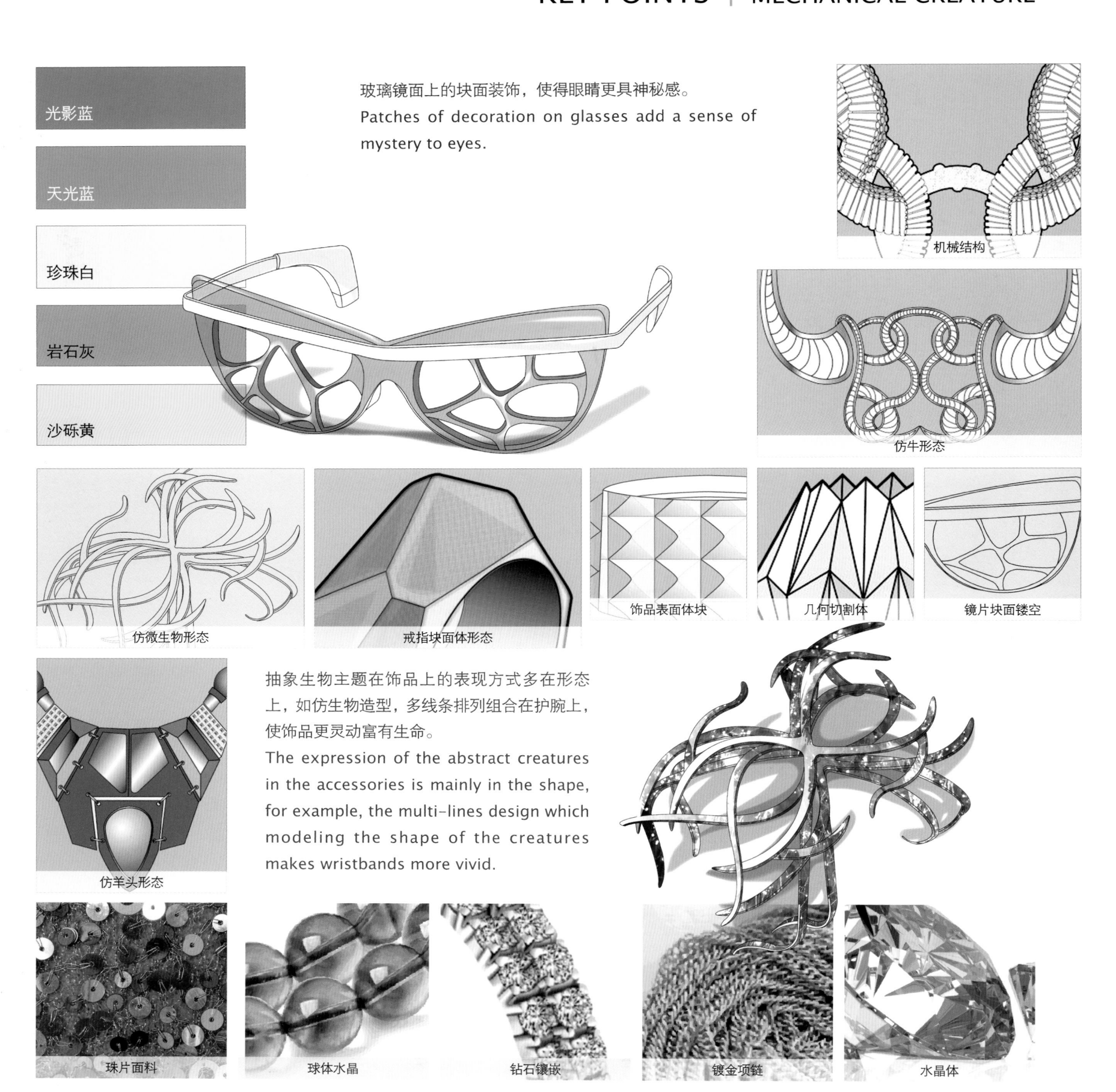

光影蓝

天光蓝

珍珠白

岩石灰

沙砾黄

机械结构

仿牛形态

仿微生物形态

戒指块面体形态

饰品表面体块

几何切割体

镜片块面镂空

仿羊头形态

珠片面料

球体水晶

钻石镶嵌

镀金项链

水晶体

157

S 帽子 2015 春夏海派时尚流行趋势

时空筑梦
REALISING DREAMS THROUGH SPACE

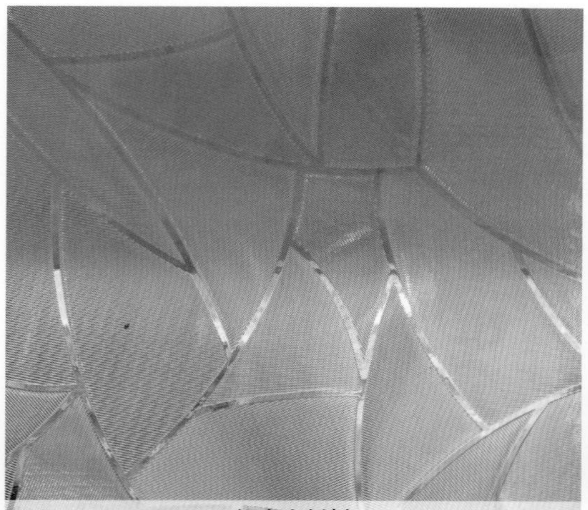

铝合金材料

PVC 条带

细致纹路

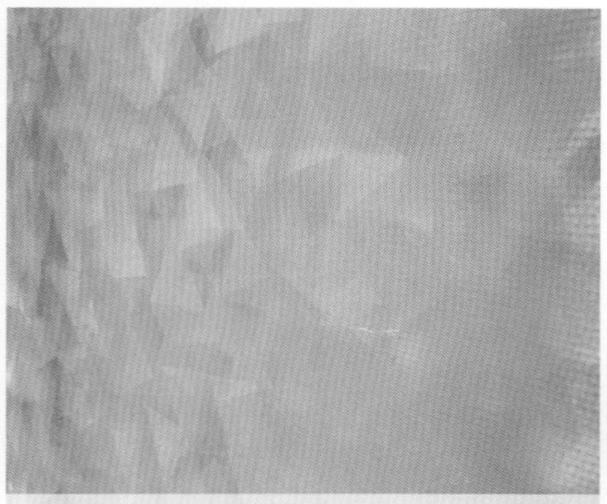

纯色材料拼接

渐变花纹

欧式翻领

马赛克拼接

设计灵感

单纯设计无需过多修饰，简单的帽檐配上细致的材质就可以表现单纯的设计。

Less is more. Well-textured fabric adds a elegant finish to this hat.

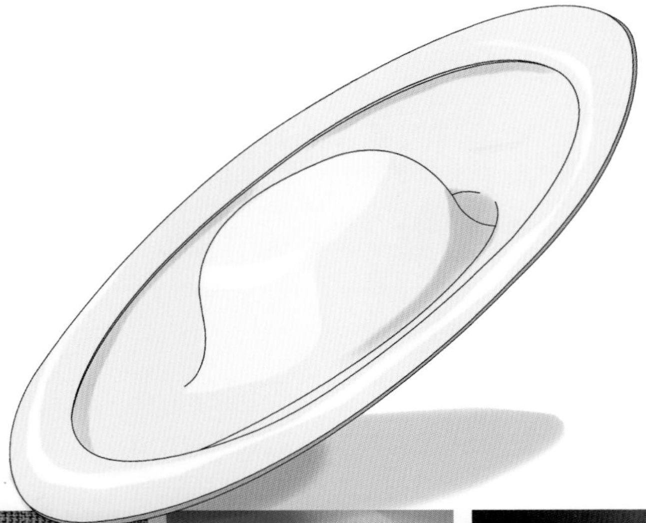

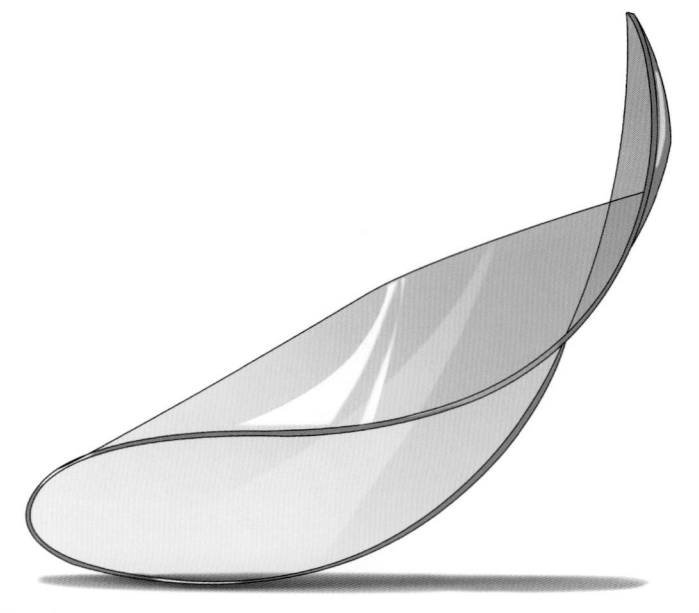

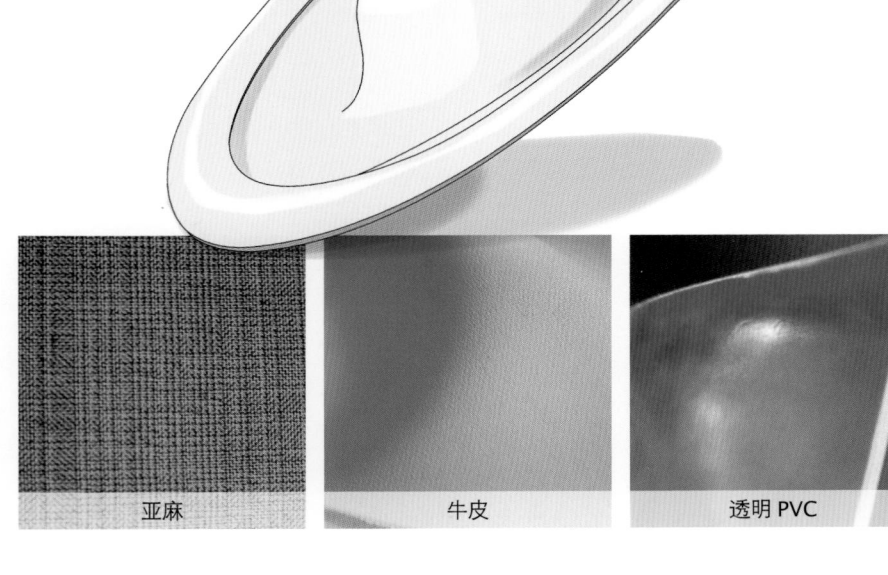

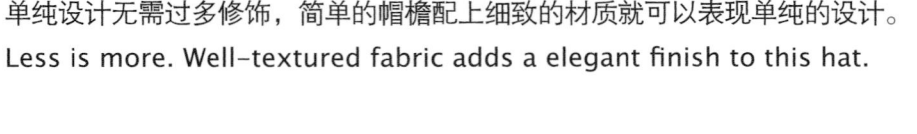

亚麻

牛皮

透明 PVC

158　www.style.sh.cn

2015 SPRING/SUMMER STYLE SHANGHAI FASHION TREND / JEWELRY

关键要素 | 单纯设计
KEY POINTS | SIMPLE DESIGN

烟云灰

氯化灰

珍珠白

日光白

天光蓝

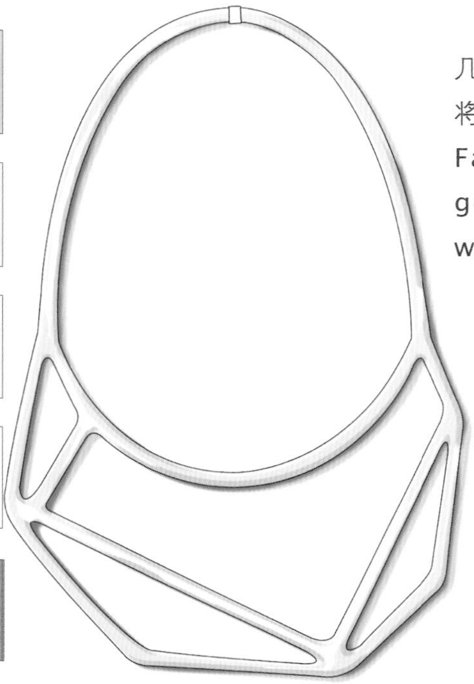

几何块面与纯白亮面材质的结合，
将会成为下一季新的设计亮点。
Fashion necklace made with
geometric patches and pure
white materials.

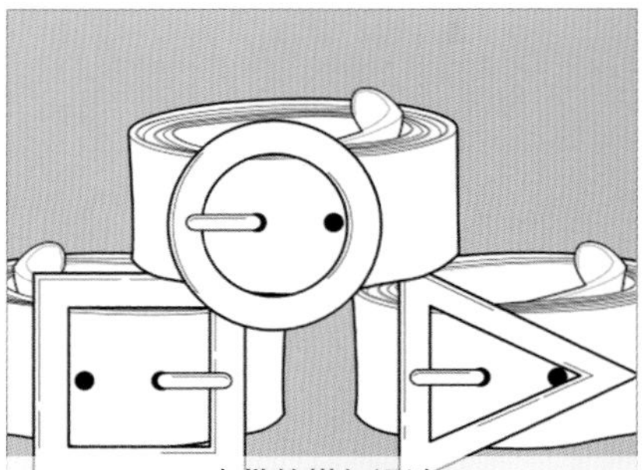

皮带的搭扣设计

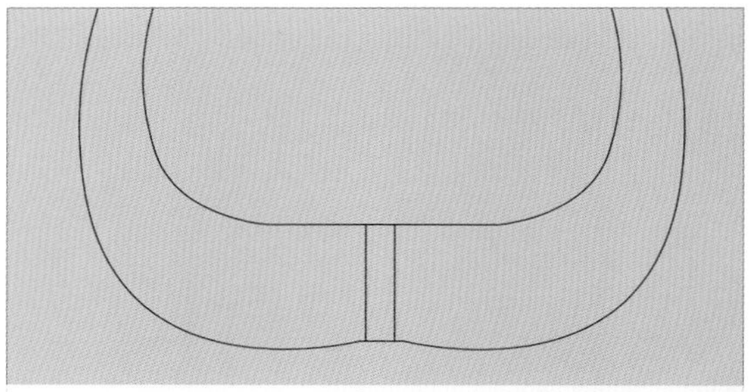

挂坠设计

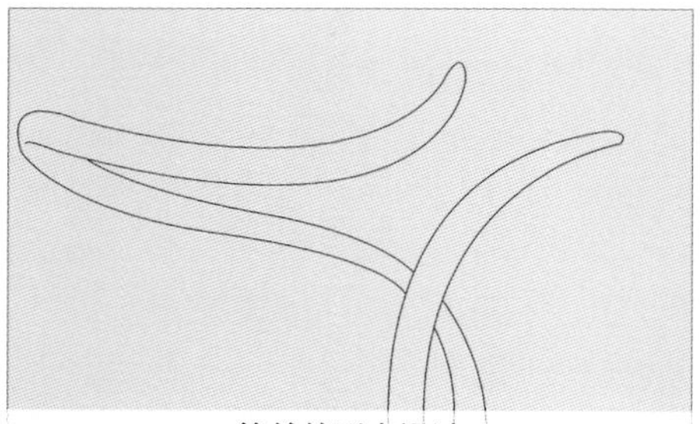

简单的形态设计

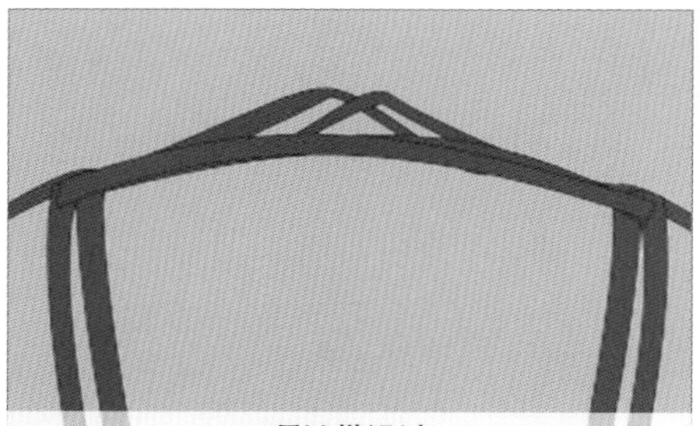

项链带设计

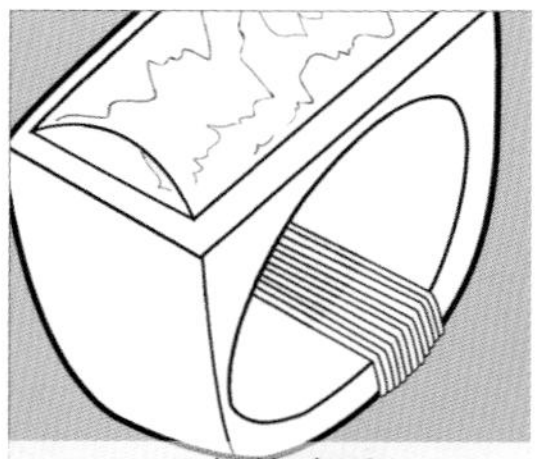

几何体造型

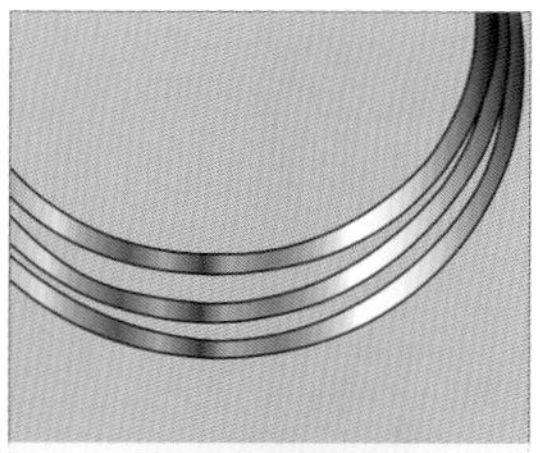

单色亮面的挂圈

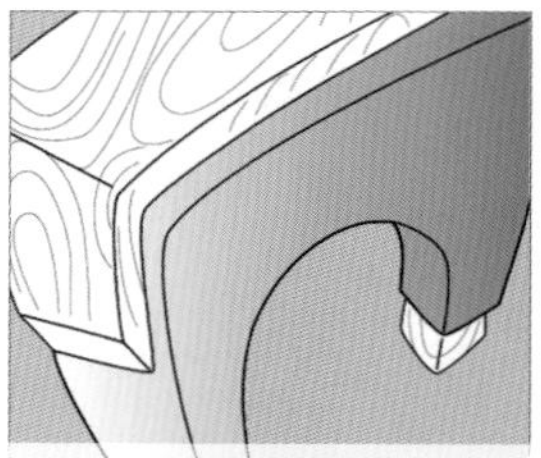

戒面上的体块

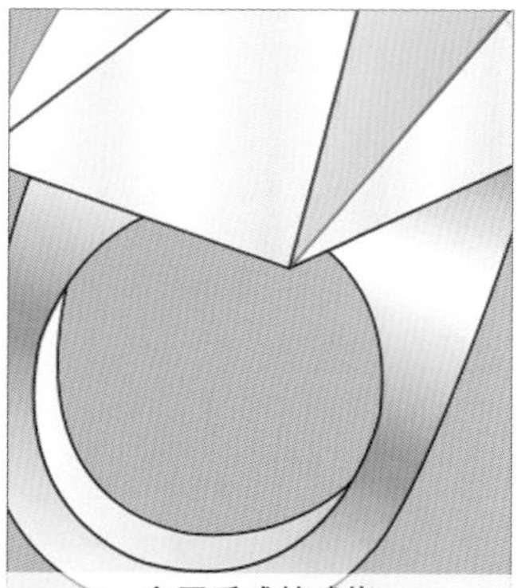

单纯无修饰的首饰品一直是时尚人的喜爱，单
一材质配上完美比例的形态，散发饰品的独特
魅力。
Beautifully designed jewelries made
of one material are essentials in
trendsetters' jewelry boxes.

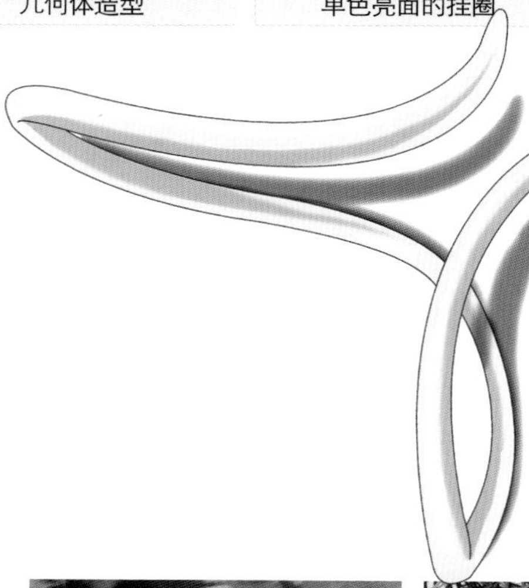

金属质感的戒指

亮面皮

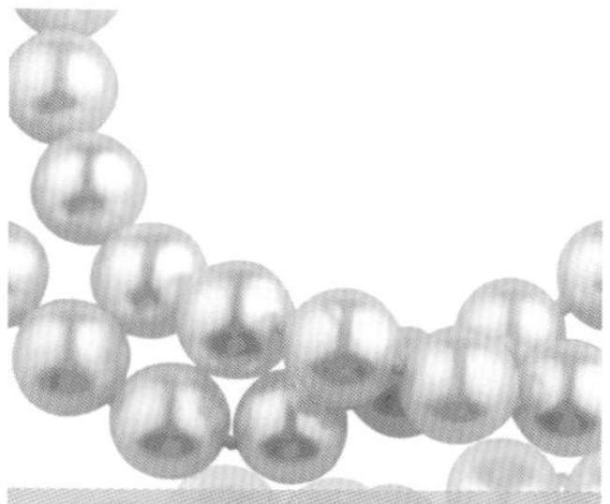

碎碟

翡翠

石榴石

金属丝

159

 配饰 2015 春夏海派时尚流行趋势

多彩享梦
ENJOY COLORFUL DREAMS

对于天马行空的梦想，像是绚丽的极光，而在心灵之外面的精彩也许是梦想生长的温床，沾着清晨的露水飘来到窗前，一种朦胧的美妙感荡漾在心中，梦想溯着月光腾飞于太空远离床前，游离的思绪让心久久不能平静，试着回想童年的记忆，回忆那一段色彩斑斓的时间，享受属于你的色彩。

Fancy dreams are like brilliant aurora, colorful but intangible. Only real world is an incubator for our dreams. When we wake up with fresh morning dew dipping on windows, we still have some hazy memories about what we just dreamt. When we lie on bed every night, the Sandman comes and brings us to another dreamland above universe. Dreams can help us exercise the power of imagination. Try to recall sweet childhood memories and paint your world with different colors.

HAT, JEWELRY & OTHERS

2015 SPRING/SUMMER STYLE SHANGHAI FASHION TREND
海派帽子、首饰及其他流行趋势

2015 SPRING/SUMMER STYLE SHANGHAI FASHION TREND / **ACCESSORY**

关键要素 | KEY POINTS

童年拾趣
CHILDHOOD REFLECTIONS

跨界实验
TRANSBOUNDARY EXPERIMENTS

快乐崇拜
JOLLY IDOLATRY

都市街头
CITY STREETS

浅裸粉

蜜桃粉

幻彩红

冰山蓝

玛瑙蓝

松石绿

珊瑚红

迎春黄

琵琶橙

石榴红

其它 2015 春夏海派时尚流行趋势

多彩享梦
ENJOY COLORFUL DREAMS

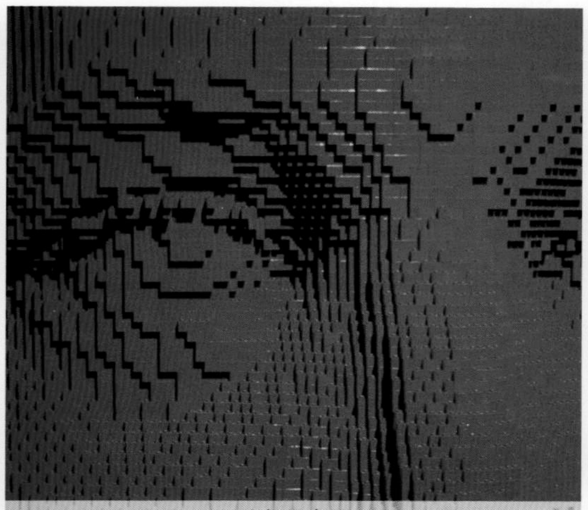

乐高积木展

彩色乐高堆叠

装饰艺术画

彩色纽扣

彩色便签纸

趣味的公仔

装饰玻璃

设计灵感

将英文练习本的条纹线与字母结合，运用到饰品的设计中，让人联想到童年美好的记忆。

It is very interesting to combine alphabets in Italian font with watches and bracelet.

条纹织带

塑料乐高

镀银链条

162　www.style.sh.cn

2015 SPRING/SUMMER STYLE SHANGHAI FASHION TREND /HAT

关键要素 | 童年拾趣
KEY POINTS | CHILDHOOD REFLECTIONS

珊瑚红

石榴红

霓虹绿

湖泊蓝

迎春黄

在帽子的图形面料上结合童年的文具用品形象，打造时髦前卫、充满回忆的款式。

People are likely to be attracted by items that can evoke childhood memories such as stationary used many years ago. So we apply these prints to the hat.

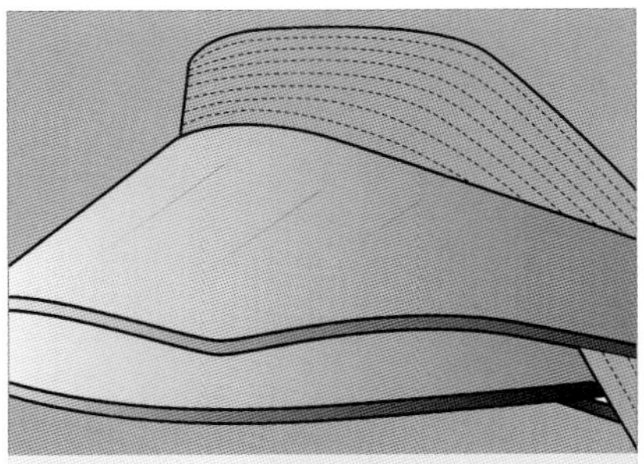

异形帽檐

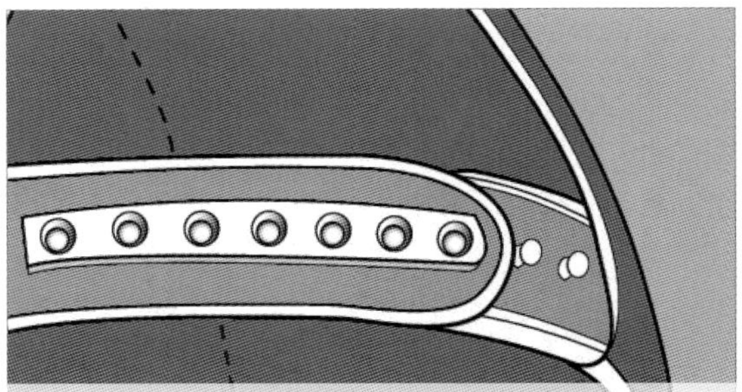

塑料可收缩尾带

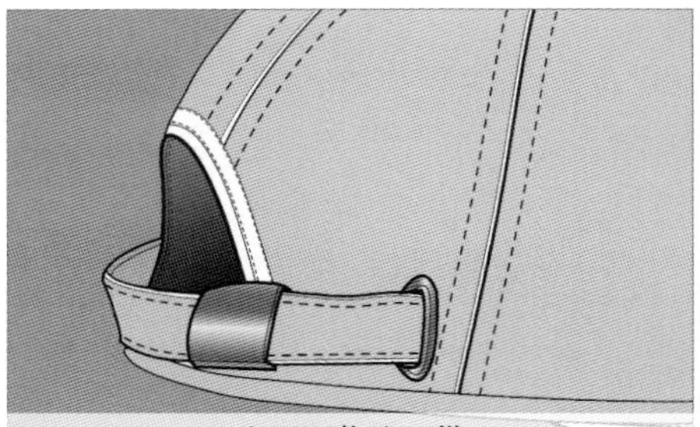

金属可收缩尾带

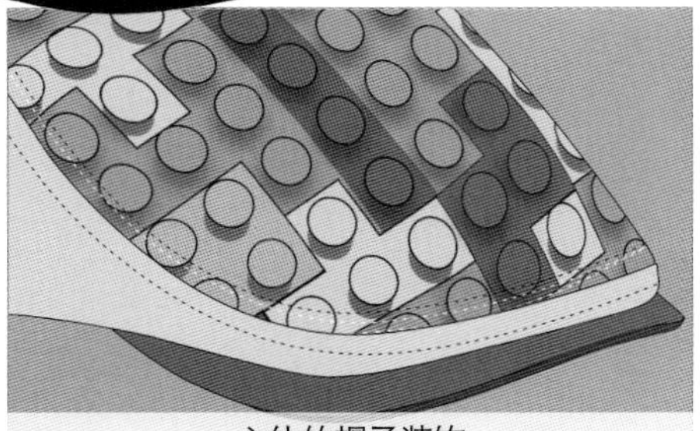

立体的帽子装饰

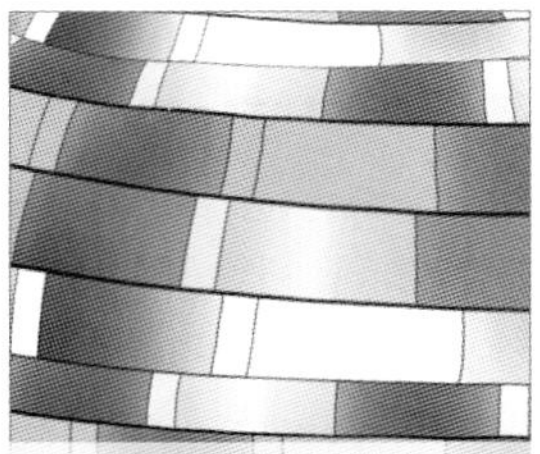

间色的组合形式

飞行棋图案

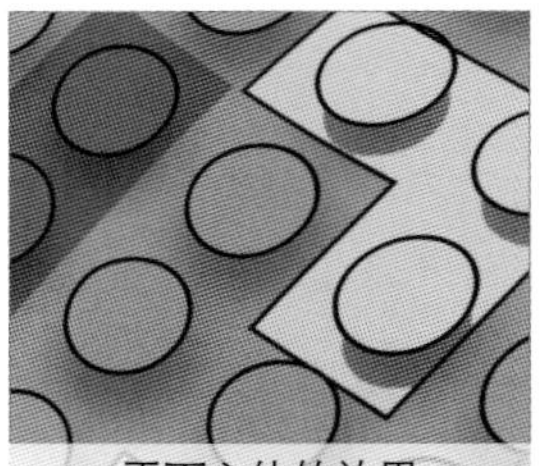

平面立体的效果

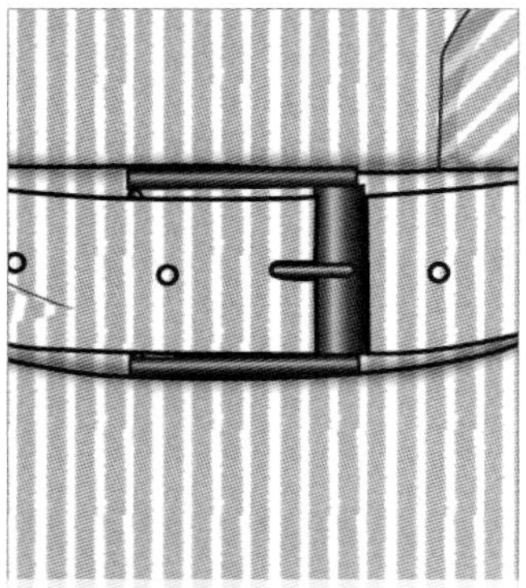

可拆卸的腰条

主要通过飞行棋的元素变形后运用在帽子的各个部分。

Variation of elements in flight chess brightens the hat.

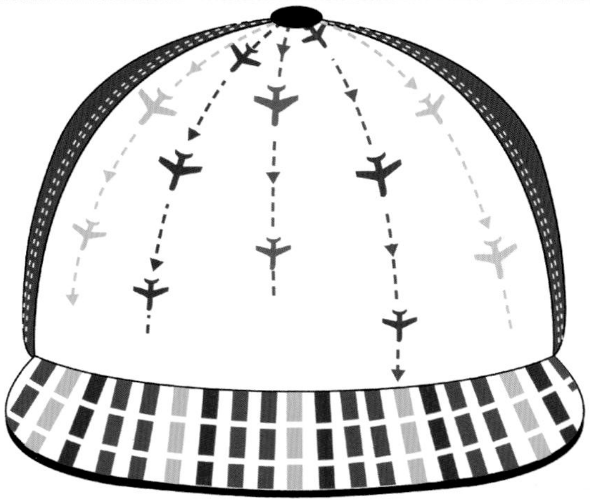

尼龙绳

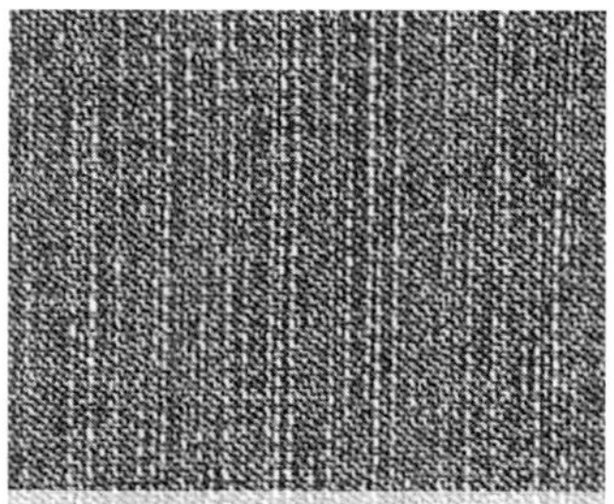

纯色牛仔布

格纹布

编织皮绳

银光粉 PVC

163

首饰 2015 春夏海派时尚流行趋势

多彩享梦
ENJOY COLORFUL DREAMS

点线面的构成 　　　平面几何构成 　　　广告插图

变异的装饰画　　　发散性透视 　　　广告装饰艺术　　　渐变效果

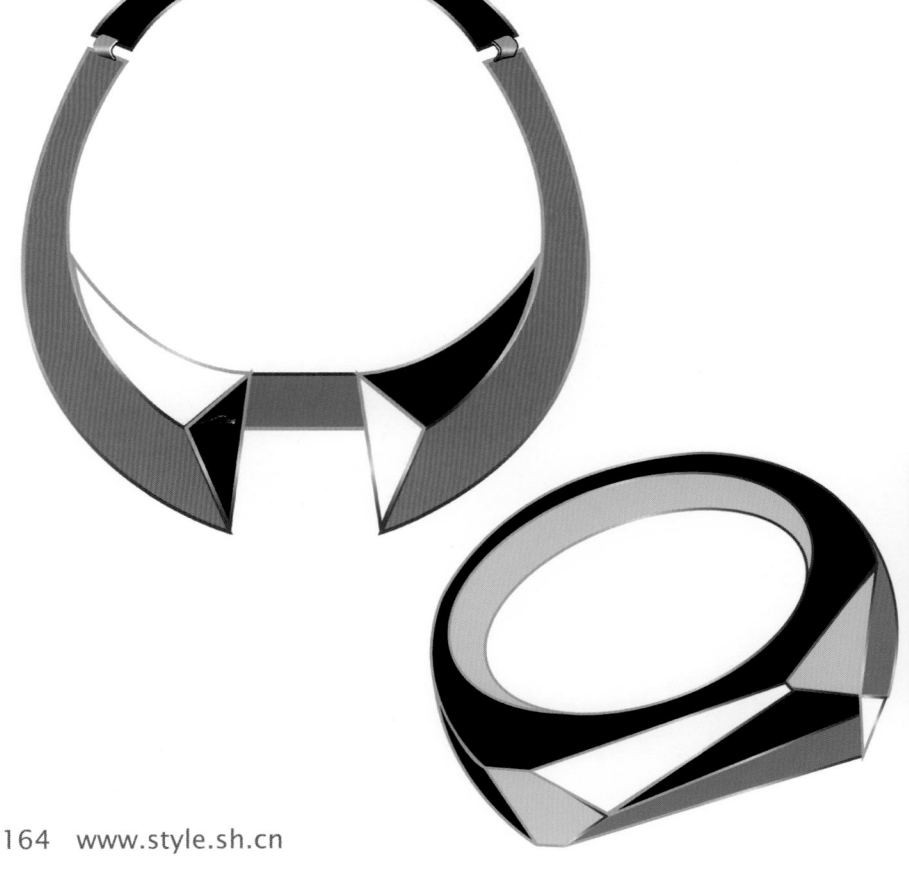

设计灵感

不规则的块面分割，让首饰打破常规，加上色块上的强烈对比，让跨界的首饰不同寻常。

Irregular shaped surfaces in contrasting colors make these design very outstanding.

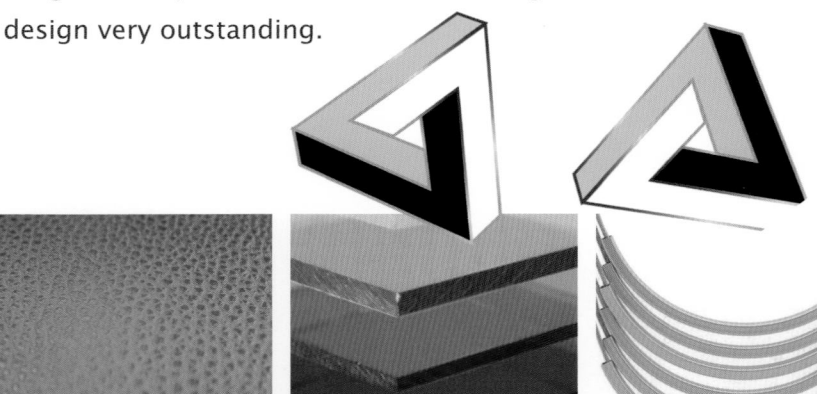

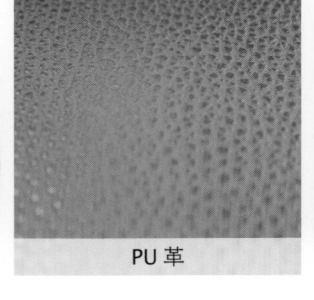

PU 革　　　彩色亚克力　　　镀金

164　www.style.sh.cn

2015 SPRING/SUMMER STYLE SHANGHAI FASHION TREND / HAT

关键要素 | 跨界实验
KEY POINTS | TRANSBOUNDARY EXPERIMENTS

湖泊蓝

浅草绿

迎春黄

幻彩红

松石绿

高高的帽顶搭配飞行眼镜造型，仿佛透过 PVC 镜片透出炯炯有神的两个眼睛。
Under the large trim is a pair of PVC glasses. This design highlights the eyes.

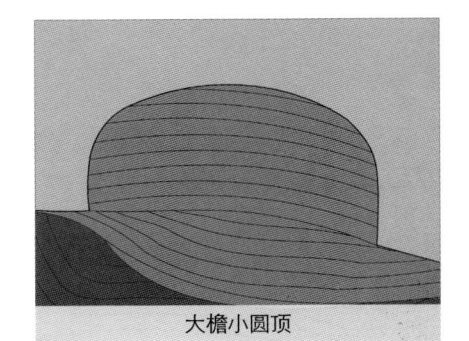
大檐小圆顶

编织腰条对比色装饰

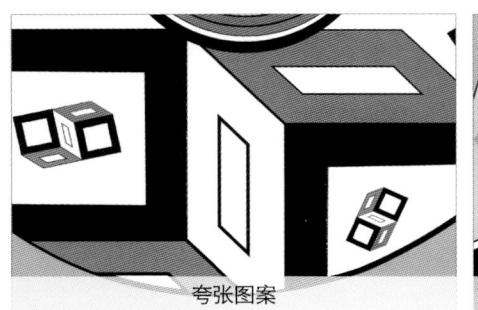
夸张图案

平面印刷造成的视错觉

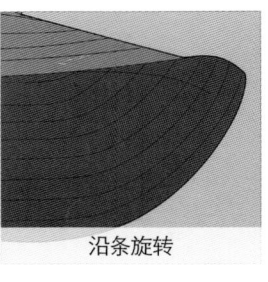
沿条旋转

立体形态图案

彩色沿条拼接

附加眼镜

跨界地把不同的风格、图形、材质结合在一起，大胆的图形为传统的帽子注入新鲜活力。
Different styles, graphic designs and materials add vitality to classic hats.

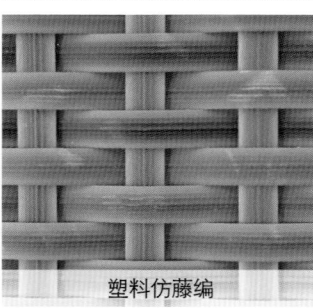
塑料仿藤编

编织便条

牛津布

印花牛仔布

合金

165

 帽子 2015 春夏海派时尚流行趋势

多彩享梦
ENJOY COLORFUL DREAMS

LED 灯下的外白渡桥

彩色糖果

哑光镭射光泽

清澈的海滩

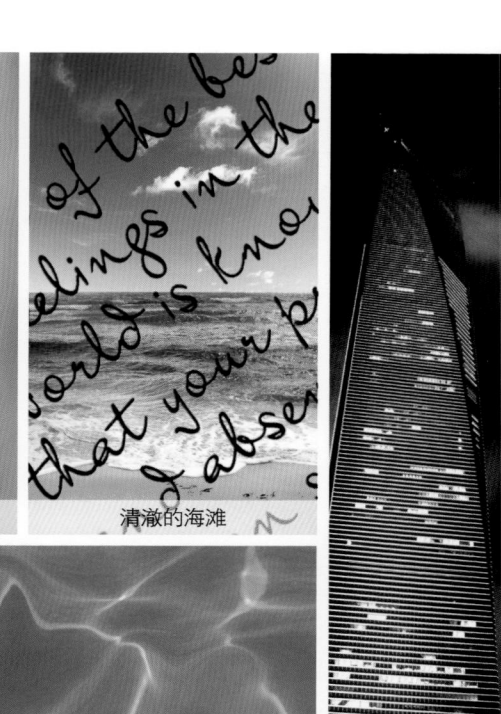

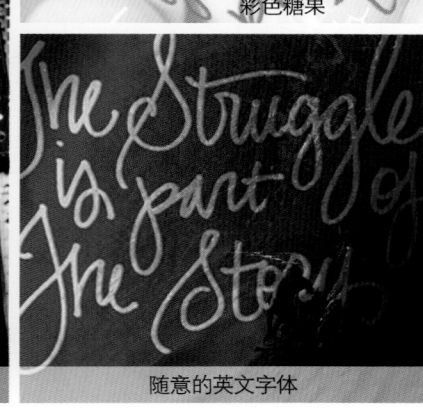

随意的英文字体

水波纹

LED 包裹的高

设计灵感

在帽子上将塑料、花纹帆布与皮革混合形成对比，透明的着色塑料别具一格，手写流畅的人工文字和天然的自然风景放在一起打造一种新意，更高的帽顶改良轮廓，打造时髦的都市学院风造型，快乐自由。

This hat combines plastic, floral printed canvas and leather. Smooth letters and landscape prints lend novelty to it. The high shape hat shaping the modern , free, preppy look.

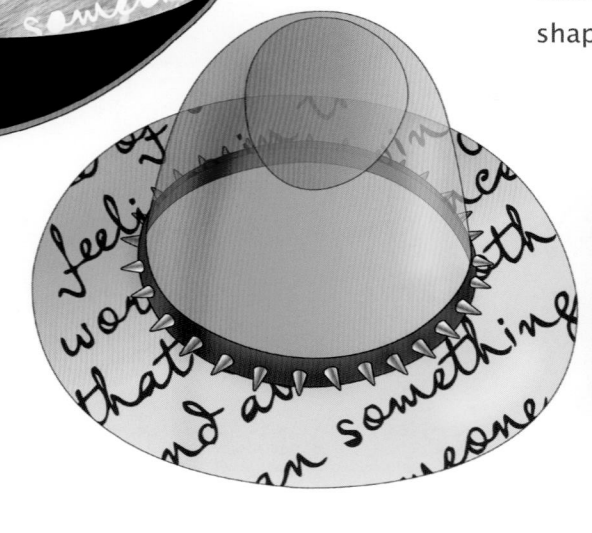

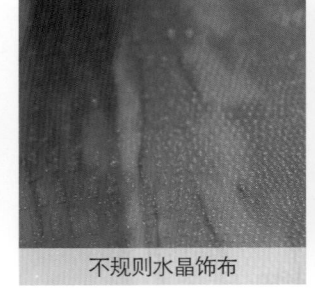

不规则水晶饰布

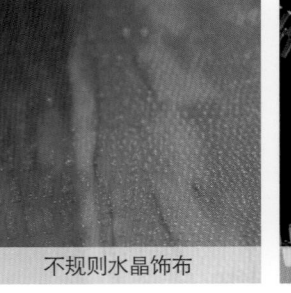

金属拉链

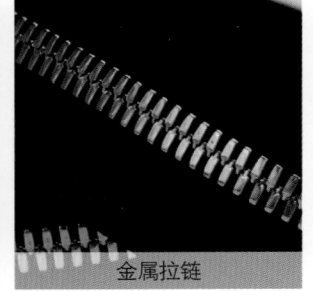

彩色烤漆铆钉

2015 SPRING/SUMMER STYLE SHANGHAI FASHION TREND / OTHERS

关键要素 | 快乐崇拜
KEY POINTS | JOLLY IDOLATRY

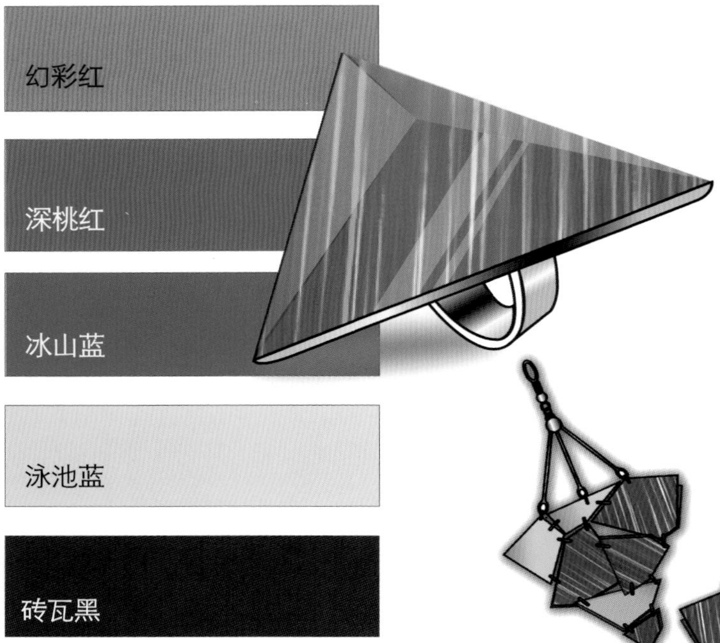

幻彩红

深桃红

冰山蓝

泳池蓝

砖瓦黑

随意的块面分割和彩色线状的印花，更为耳环增添几分灵动。
Irregular blocks and colorful line prints style for a playfully eclectic look.

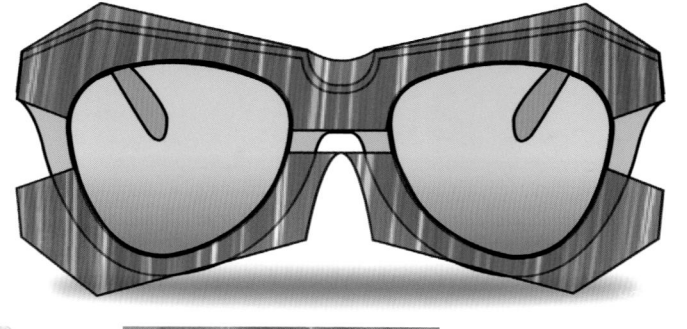

大镜框

外发光线性肌理

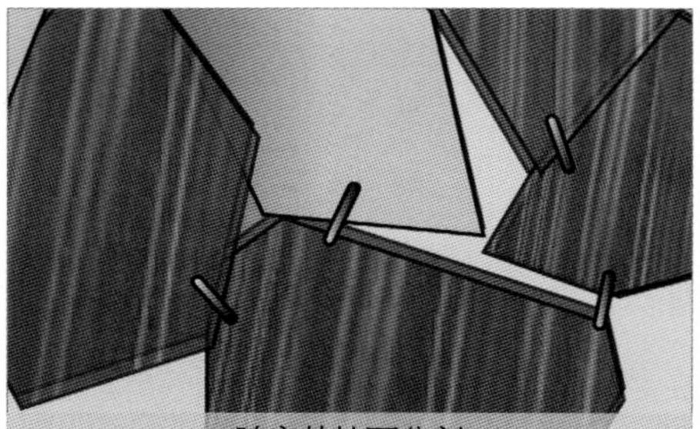

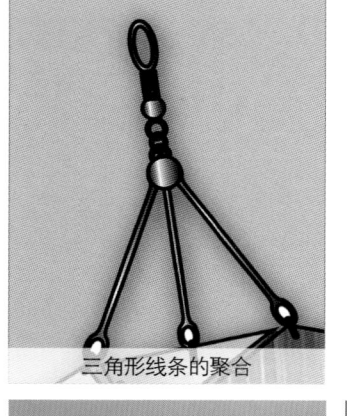

随意的块面分割

粗线条

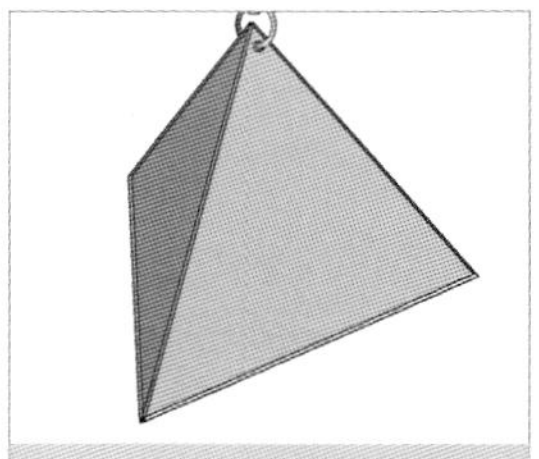

几何透视感

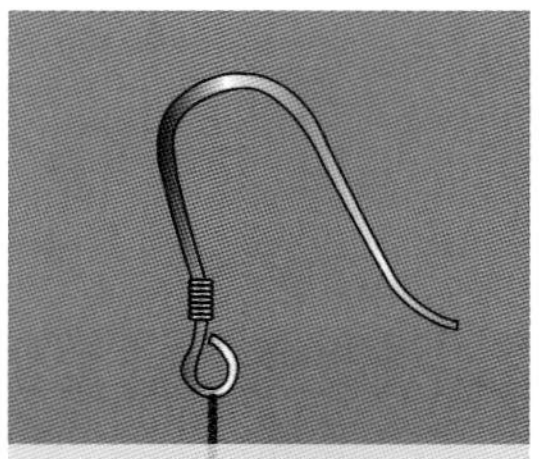

挂耳式设计

硬朗的块面分割

夸张的眼镜廓型加上通透的材质，强烈的层次感中透出几分趣味性。
More is more on fun oversized sunglasses with translucent frames and contrast temples.

三角形线条的聚合

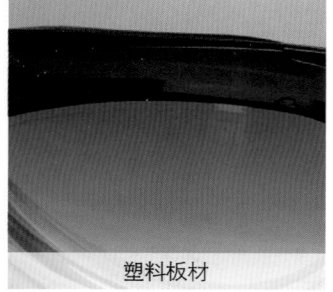

塑料板材

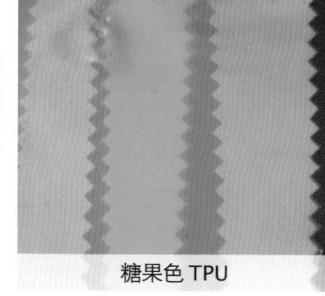

糖果色 TPU

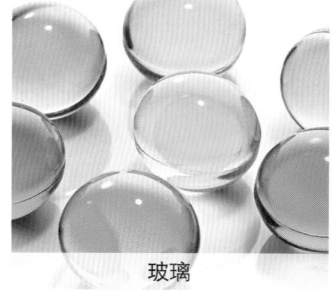

玻璃

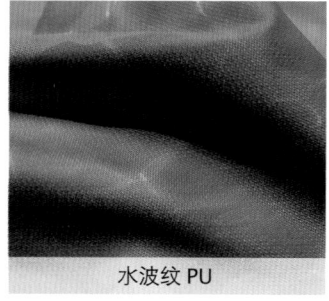

水波纹 PU

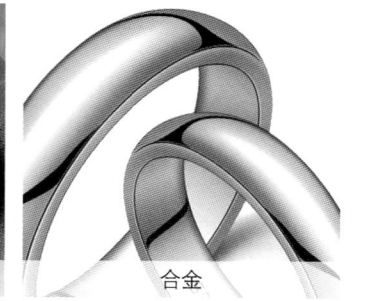

合金

167

S 帽子 2015 春夏海派时尚流行趋势

多彩享梦
ENJOY COLORFUL DREAMS

卡通漫画图像设计

用丁字裤做的构成展示作品

涂鸦艺术

波普艺术处理金刚形象

面料肌理转化成图形贴在车体上

儿童卡通用品店外观

街头咖啡吧

设计灵感

棒球帽和骑士帽的图案采用鲜艳亮丽的涂鸦形式表现出来，或简约低调的帽檐翻出的里布显现出来街头的感觉。

Graffiti style prints on horseman hat and cap and street style colorful interior.

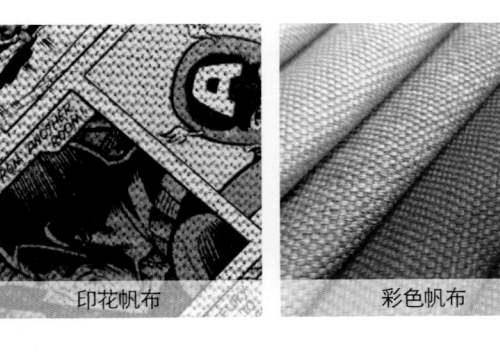

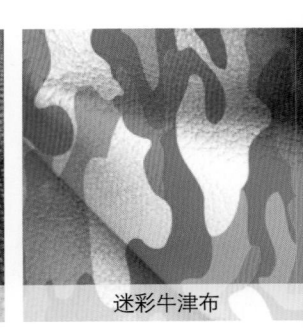

印花帆布

彩色帆布

迷彩牛津布

168 www.style.sh.cn

2015 SPRING/SUMMER STYLE SHANGHAI FASHION TREND / OTHERS

关键要素 | 都市街头
KEY POINTS | CITY STREETS

乌梅黑

湖泊蓝

石榴红

金币黄

松石绿

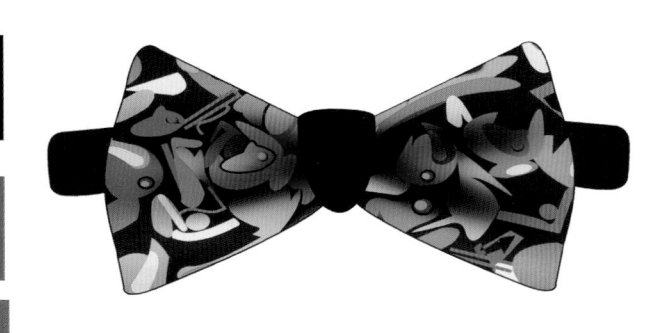

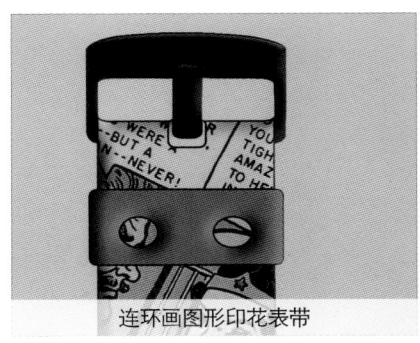

连环画图形印花表带

重复缠绕的彩色皮手链，附上报纸图形的印花，街头感十足。
The multi-wrap colorful leather strap of a bracelet makes a chic statement with a touch of newspaper prints.

缠绕的皮手链

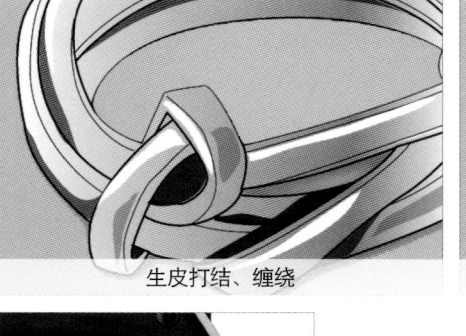

生皮打结、缠绕

大尺寸的表面

迷彩元素

上扬的眼镜架

涂鸦印花

粗犷的表面廓型

在街头的墨镜中注入漫画元素，表现随性的感觉。
The multi-wrap colorful leather strap of a bracelet makes a chic statement with a touch of newspaper prints.

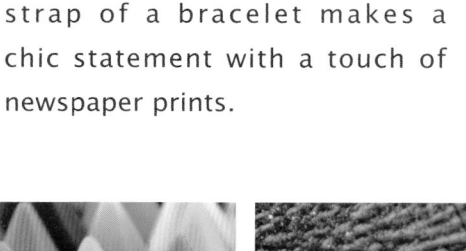

塑料板材

彩色烤漆铆钉

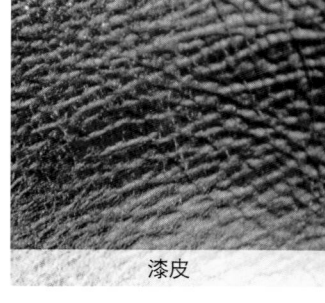

漆皮

牛仔布

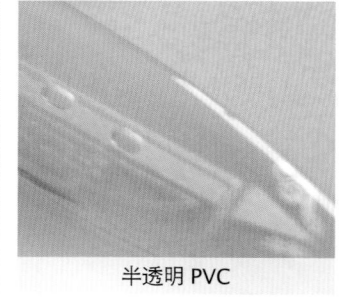

半透明 PVC

169

2015 春夏海派时尚流行要素汇编·配饰篇
2015 SPRING/SUMMER STYLE SHANGHAI FASHION ELEMENTS COLLECTION·ACCESSORY

海上星梦
STAR DREAMS OVER THE SEA

色彩

浪花白	青釉灰	青砖灰	乌梅黑	麦糖黄	胭脂红	枫叶红	紫檀棕
SH1200318	SH0801254	SH1101076	SH1000211	SH0101301	SH0303138	SH0303471	SH0303477

皮革黄	杏仁黄	琥珀橙	酸枝红	瓷瓦黑	雷雨灰	铁艺灰	青瓷灰
SH0101365	SH0101055	SH0200513	SH0303323	SH1100911	SH1100175	SH1100994	SH1100980

鞋履

繁华时代	精细品质	新贵气质	创新传统
FLOURISHING ERA	EXQUISITE QUALITY	NOBLE TEMPERAMENT	INNOVATING TRADITIONS

箱包

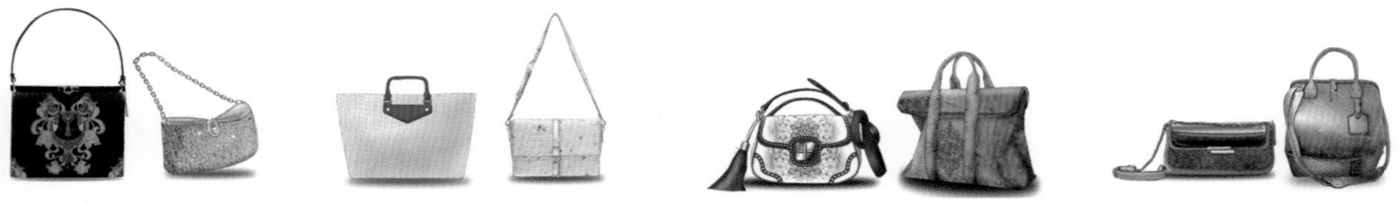

繁华时代	精细品质	新贵气质	创新传统
FLOURISHING ERA	EXQUISITE QUALITY	NOBLE TEMPERAMENT	INNOVATING TRADITIONS

帽子

繁华时代	精细品质	新贵气质	创新传统
FLOURISHING ERA	EXQUISITE QUALITY	NOBLE TEMPERAMENT	INNOVATING TRADITIONS

饰品

繁华时代	精细品质	新贵气质	创新传统
FLOURISHING ERA	EXQUISITE QUALITY	NOBLE TEMPERAMENT	INNOVATING TRADITIONS

优活之梦
LIVING THE DREAM

色彩

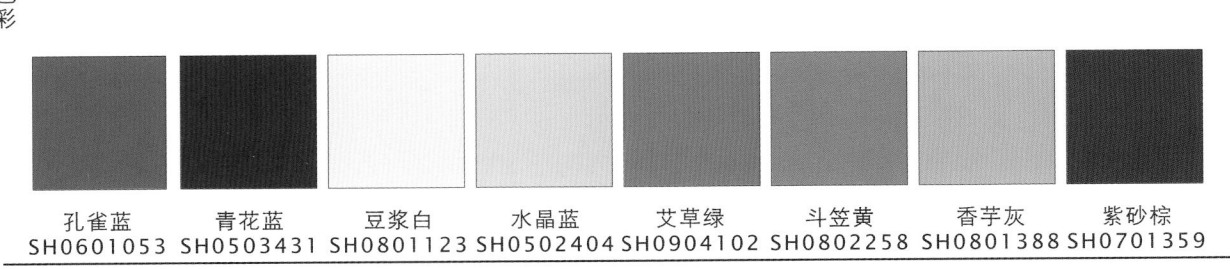

孔雀蓝	青花蓝	豆浆白	水晶蓝	艾草绿	斗笠黄	香芋灰	紫砂棕
SH0601053	SH0503431	SH0801123	SH0502404	SH0904102	SH0802258	SH0801388	SH0701359

苔藓绿	蜜瓜黄	枣泥红	剑麻灰	抹茶绿	淡奶黄	蛋壳黄	曜石黑
SH0903166	SH0100139	SH0301069	SH0802256	SH0901249	SH0100230	SH0201527	SH1000027

鞋履

织补风尚	平民贵族	生态材质	家庭手工
BACK TO THE BASICS	ORDINARY NOBILITY	ECO MATERIAL	FAMILY HANDICRAFTS

箱包

织补风尚	平民贵族	生态材质	家庭手工
BACK TO THE BASICS	ORDINARY NOBILITY	ECO MATERIAL	FAMILY HANDICRAFTS

帽子

织补风尚	平民贵族	生态材质	家庭手工
BACK TO THE BASICS	ORDINARY NOBILITY	ECO MATERIAL	FAMILY HANDICRAFTS

饰品

织补风尚	平民贵族	生态材质	家庭手工
BACK TO THE BASICS	ORDINARY NOBILITY	ECO MATERIAL	FAMILY HANDICRAFTS

时空筑梦
REALISING DREAMS THROUGH SPACE

色彩

氯化灰	珍珠白	光影蓝	亚铜灰	柠檬黄	林荫绿	碳钢灰	天光蓝
SH0800262	SH0800210	SH0500285	SH0800481	SH0101109	SH0906065	SH1100763	SH0602025

日光白	陨石棕	恒星橙	烟云灰	苹果绿	沙砾黄	象牙黄	岩石灰
SH1200317	SH0701599	SH0201888	SH1100987	SH0900548	SH0800188	SH0801193	SH0800456

鞋履

复制粘贴	符号人生	机械生物	单纯设计
REPETITION & MODULIZATION	SYMBOLIC LIFE	MECHANICAL CREATURE	SIMPLE DESIGN

箱包

复制粘贴	符号人生	机械生物	单纯设计
REPETITION & MODULIZATION	SYMBOLIC LIFE	MECHANICAL CREATURE	SIMPLE DESIGN

帽子

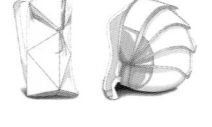

复制粘贴	符号人生	机械生物	单纯设计
REPETITION & MODULIZATION	SYMBOLIC LIFE	MECHANICAL CREATURE	SIMPLE DESIGN

饰品

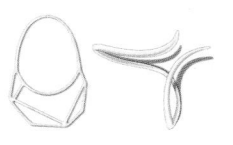

复制粘贴	符号人生	机械生物	单纯设计
REPETITION & MODULIZATION	SYMBOLIC LIFE	MECHANICAL CREATURE	SIMPLE DESIGN

172　www.style.sh.cn

多彩享梦

ENJOY COLORFUL DREAMS

色彩

蜜桃粉	浅裸粉	幻彩红	深桃红	冰山蓝	泳池蓝	玛瑙蓝	松石绿
SH0402308	SH0801319	SH0300320	SH0402644	SH0602026	SH0503462	SH0600401	SH0601123

珊瑚红	迎春黄	琵琶橙	浅草绿	湖泊蓝	石榴红	霓虹绿	金币黄
SH0202768	SH0101222	SH0101118	SH0904042	SH0600419	SH0301062	SH0904632	SH0100081

鞋履

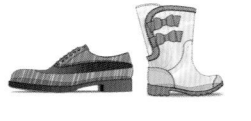

童年拾趣	跨界实验	快乐崇拜	都市街头
CHILDHOOD REFLECTIONS	TRANSBOUNDARY EXPERIMENTS	JOLLY IDOLATRY	CITY STREETS

箱包

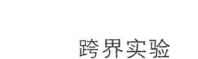

童年拾趣	跨界实验	快乐崇拜	都市街头
CHILDHOOD REFLECTIONS	TRANSBOUNDARY EXPERIMENTS	JOLLY IDOLATRY	CITY STREETS

帽子

童年拾趣	跨界实验	快乐崇拜	都市街头
CHILDHOOD REFLECTIONS	TRANSBOUNDARY EXPERIMENTS	JOLLY IDOLATRY	CITY STREETS

饰品

童年拾趣	跨界实验	快乐崇拜	都市街头
CHILDHOOD REFLECTIONS	TRANSBOUNDARY EXPERIMENTS	JOLLY IDOLATRY	CITY STREETS

图书在版编目（CIP）数据

海派时尚：2015 春夏海派时尚流行趋势．配饰篇 /
海派时尚流行趋势研究中心著 . -- 上海：东华大学出版
社，2013.10
ISBN 978-7-5669-0383-9
Ⅰ．①海… Ⅱ．①海… Ⅲ．①服饰 - 市场预测 -
上海市 -2015 Ⅳ．① TS941.13
中国版本图书馆 CIP 数据核字（2013）第 248245 号

责任编辑 杜亚玲
封面设计 范乃文 胡弘毅 周 吉

海派时尚：2015 春夏海派时尚流行趋势·配饰篇
Haipai Shishang: 2015 Chunxia Haipai Shishang Liuxing Qushi·Peishipian
海派时尚流行趋势研究中心 著

出 版：东华大学出版社（上海市延安西路 1882 号 邮政编码：200051）
出版社网址：http://www.dhupress.net
天猫旗舰店：http://dhdx.tmall.com
营销中心：021-62193056 62373056 62379558
印 刷：上海中华商务联合印刷有限公司
开 本：787mm×1092mm 1/8 印张：22
字 数：578 千字
版 次：2013 年 10 月第 1 版
印 次：2013 年 10 月第 1 次印刷
书 号：ISBN 978-7-5669-0383-9/TS•448
定 价：580.00 元